Advanced Methods in Molecular Biology and Biotechnology

Advanced Methods in Molecular Biology and Biotechnology

A Practical Lab Manual

Khalid Z. Masoodi

Transcriptomics Laboratory, Division of Plant Biotechnology, Sher-e-Kashmir University of Agricultural Sciences and Technology of Kashmir, Srinagar, India

Sameena Maqbool Lone

Division of Vegetable Science, Sher-e-Kashmir University of Agricultural Sciences and Technology of Kashmir, Shalimar, Srinagar, India

Rovidha Saba Rasool

Division of Plant Pathology, Sher-e-Kashmir University of Agricultural Sciences and Technology of Kashmir, Shalimar, Srinagar, India

ELSEVIER

ACADEMIC PRESS

An imprint of Elsevier

Academic Press is an imprint of Elsevier
125 London Wall, London EC2Y 5AS, United Kingdom
525 B Street, Suite 1650, San Diego, CA 92101, United States
50 Hampshire Street, 5th Floor, Cambridge, MA 02139, United States
The Boulevard, Langford Lane, Kidlington, Oxford OX5 1GB, United Kingdom

Notices
Knowledge and best practice in this field are constantly changing. As new research and experience
broaden our understanding, changes in research methods, professional practices, or medical treatment
may become necessary.

Practitioners and researchers must always rely on their own experience and knowledge in evaluating
and using any information, methods, compounds, or experiments described herein. In using such
information or methods they should be mindful of their own safety and the safety of others, including
parties for whom they have a professional responsibility.

To the fullest extent of the law, neither the Publisher nor the authors, contributors, or editors, assume
any liability for any injury and/or damage to persons or property as a matter of products liability,
negligence or otherwise, or from any use or operation of any methods, products, instructions, or ideas
contained in the material herein.

Library of Congress Cataloging-in-Publication Data
A catalog record for this book is available from the Library of Congress

British Library Cataloguing-in-Publication Data
A catalogue record for this book is available from the British Library

ISBN: 978-0-12-824449-4

For information on all Academic Press publications
visit our website at https://www.elsevier.com/books-and-journals

Publisher: Andre Gerhard Wolff
Acquisitions Editor: Peter Linsley
Editorial Project Manager: Mona Zahir
Production Project Manager: Sreejith Viswanathan
Cover designer: Greg Harris

Typeset by SPi Global, India

Special thanks to
Mr. Parvez Ahmad Bhat
Deputy Registrar (Academics), SKUAST-Kashmir for his Constant
Guidance, Encouragement, and Moral Support.

Contents

Preface

Amendments and evolution in science and technology come about through time. Innovative notions transmute our way of looking at things and give us new acumens into areas of knowledge with the advent of time, which had previously been uncertain. Revolution in molecular biology began with the discovery of DNA's double helix structure, discovery of Taq polymerase, restriction enzymes, and with the cracking of the genetic code. Ever since the advent of biotechnology, old breeding methods that took thousands of years to generate end products that can be used by humans are becoming archaic. The intensity up to which we can produce predestined genetic adjustments in bacteria, plants, and animals with high precision has grown exponentially and we are able to produce human proteins, enzymes, etc. through biotechnological interventions.

The blueprint of this lab manual is simple yet deals with experiments that are translational, with the different chapters dealing with methods and experiments of basic biotechnology and molecular biology. The principle behind each experiment is defined so that a student can understand the concepts underlining each experiment. Since we have made the individual experiments independent, students and researchers can start from any experiment. Although some experiments require knowledge of basic experiments that are covered in chapters at the beginning of this book, each experiment starts with an introduction explaining the experiment followed by the general principle of the experiment. This is followed by the materials and reagents required to conduct the experiment and the step-by-step procedure. The chemicals used in the experiment are listed in the section on reagents. The principle underlining each chemical that is used in the experiment is written with the chemical name in this section. The experiment ends with the Precautions that need to be taken care of during the experiment and Inference of the experiment at the end. A small picture of the experiment setup is also given toward the end of each experiment.

In writing this practical manual, we have tried to tone it for students at a level that would be Undergraduates, MSc, and Ph.D. students of all academic and research institutions across the world.

Khalid Z. Masoodi
Assistant Professor
Plant Biotechnology
SKUAST-Kashmir

Foreword

Nazeer Ahmed

It gives me immense pleasure to write a foreword for practical biotechnology manual entitled *"Advanced Methods in Molecular Biology and Biotechnology—A Practical Lab Manual"* written by my young colleagues. It has been another accomplishment of the authors after their successful publication of *"Ethnobotany of Himalayan Botanicals"* a comprehensive book on medicinal plants for the benefit of mankind. I have had a long association with crop improvement and biotechnology throughout my career and I am happy to see biotechnology emerging as a successful field of study and research at SKUAST-Kashmir.

Biotechnology has vast applications in medicine, health, veterinary, agriculture, horticulture, and other fields of biology. A number of molecular techniques are currently used for disease diagnosis, cloning, transgenics, DNA barcoding, management of plant diseases, molecular plant breeding, and discovery of new pathogens and genes.

This practical manual is based on the current syllabus of BSc, MSc, and PhD programs running at SKUAST-K. So, I see every reason for this manual becoming essential reading for all the students for whom practical biotechnology forms a part of the syllabus.

It is an authoritative and well-researched practical manual for biotechnology laboratory. Although there are many books published locally, I found this comprehensive work has minute details and principles of each experiment presented methodically for the understanding of students, who are at any stage of their career. It contains information in well laid-out sections and in a very precise manner. This laboratory manual is user friendly and provides the reader the ability to perform various experiments easily through the descriptions given in each experiment in this manual.

I congratulate Dr. Khalid Z. Masoodi and his students Sameena Maqbool Lone and Rovidha Saba Rasool for this great publication and I am sure, this manual will provide students and scholars with copious prospects for precipitous inculcation of concepts and principles in Advanced Methods in Molecular Biology and Biotechnology—A Practical Lab Manual.

(Nazeer Ahmed)

Acknowledgment

My sincere and heartfelt thanks to SERB, Government of India for their financial support under SERB/2016/ECR0000025 and RCFC North-II, NMPB, Ministry of AYUSH, GoI under AU/FoA/RcFc/2019/445-48.

Introduction

In the field of biotechnology and molecular biology, research has progressed at a fascinating rate in recent years. Much of this progress results from the development of new laboratory techniques that allow very precise fractionation and analysis of nucleic acids and proteins, as well as the construction of recombinant DNA molecules that can then be cloned and expressed in host cells. With the passage of time, new techniques have been regularly introduced and the sensitivity of older techniques greatly improved upon.

The purpose of this book therefore is to introduce the readers to a selection of the more advanced analytical and preparative techniques, which the authors consider to be frequently used by researchers in the field of biotechnology and molecular biology. In choosing techniques for this book, we have obviously had to be very selective, and for the sake of brevity a knowledge of certain basic biochemical techniques and terminology has been assumed. It has been seen that while performing the experiments, students usually lag behind in knowing the actual background such as the principle and applications of the method performed. Hence, in choosing these chapters we have considered all these aspects and observations.

In this manual, each chapter aims to describe the theory and the relevant practical details for a given technique, and to identify both the potential and limitations of the technique. Each chapter is written by authors who regularly use the technique in their own laboratories. Step-by-step protocols and practical notes are provided. The distinguishing feature of this manual compared to others is the detailed explanation of the theoretical mechanisms of each step and the discussion of the importance of the techniques. In addition, each protocol includes an indication of the time involved in its application. This knowledge will enable the users to design their own modifications or to adapt the method to different systems. Also, this laboratory manual gives a thorough introduction to modern, cost-efficient techniques in molecular biology, which can find applications in many different fields. It is the result of practical experience, with each protocol having been used extensively in undergraduate courses and workshops or tested in the author's laboratory. Thus, they are also likely to work in inexperienced hands the first time they are performed. In writing this practical manual, we have tried to tone it down for students at a level that would be understandable by undergraduates through UG, MSc, and up to PhD students as well in the universities across India and the world as these are the stages at which we have experience of conducting these experiments. The course matter followed is that of the Indian Council of Agricultural Research (ICAR) syllabus.

About the author

Dr. Khalid Zaffar Masoodi is a young scientist working as an assistant professor at the Division of Plant Biotechnology in Sher-e Kashmir University of Agricultural Sciences and Technology of Kashmir (India). He holds a Masters and PhD in Biotechnology. After completing his PhD, he worked in Prof. Zhou Wang's Lab at University of Pittsburgh, PA, USA as postdoctoral associate (2010–14). Dr. Masoodi is a recipient of SERB Early Career Research Award 2016, Travel Award, SBUR, USA 2012, Society of Endocrinology Journal Award 2018, and many more. With a glorious academic career, Dr. Masoodi holds a distinction in qualifying CSIR-UGC NET in Life Sciences, DBT-BET JRF, and has qualified the Graduate Aptitude Test in Engineering and Sciences (GATE) while he was doing his Masters in Plant Biotecholology. His lab (Transcriptomics Laboratory, also called as K-Lab inspired by Klenow fragment that revolutionized biotechnology research) at SKUAST-K is well equipped with fluorescence imaging facility, RT-PCR, cell culture, Western blotting, and many state-of-the-art instruments. His research focus is on multidimensional work on transcriptomics of stress biology (biotic/abiotic) in plants and cancer biology. Apart from that, he is extensively working on drug discovery from medicinal plants. He has produced high-impact papers published in Oncogene, Molecular Cancer Therapeutics, Endocrinology, Journal of Functional Foods, PloS One, and many more, and has more than 230 sequences in Genbank. He also deciphered the whole transcriptome of six Kashmiri apple varieties. Currently, he has three externally funded projects from SERB, DST, which focus on high-throughput screening of medicinal plants for prostate cancer and cisgenic apples. In 2019–20 Dr. Masoodi filed five patent applications with the Indian Patent Office.

Sameena Maqbool Lone, currently, an MSc scholar in the Division of Vegetable Science, SKUAST-K, did her Bachelor's in Horticulture, from SKUAST-K, Shalimar. She is working on the Heterosis and combining ability studies in cherry tomato (*Solanum* spp.) and various technologies related to plant tissue culture under the coguidance of Dr. Khalid Z. Masoodi and is also holding an excellent academic and extracurricular record. She was also awarded merit scholarship by SKUAST-K during her Bachelor's degree program.

Rovidha Saba Rasool did her Bachelors in Agriculture from SKUAST-K, Wadura followed by Masters in Plant Pathology from SKUAST-Kashmir, Shalimar. Currently, she is pursuing her PhD from the Division of Plant Pathology, SKUAST-Kashmir. She has done commendable work in finding new genes involved in the pathogenesis of shot-hole disease of stone fruits through insertional mutagenesis in *Thyrostroma carpophilum*. She has been a good student holding an excellent academic and extracurricular record and was awarded PhD merit scholarship by SKUAST-K in 2018–19. Apart from this, Rovidha has taken part in various national-level sports competitions and has won many national and state-level awards in baseball and handball.

Disclaimer

The contents of the book provide various techniques in molecular biology and biotechnology and is intended for teachers, researchers, and students who wish to conduct experiments on the subject in their laboratories. Neither the authors nor the publisher can be held responsible for the accuracy of the information itself or the consequences from the use or misuse of the information in this book. The resources are not scrutinized but the experiments in this book have been conducted by experts, but it is the researcher's responsibility to ensure the accuracy of the information cited in this lab manual. It is reiterated for the sake of the readers of this manual that the information and the experiments presented in this lab manual is subject to change as research is a continuous process and variations in results might occur due to experimental or individual errors. Although, every attempt is made to minimize errors, there may be inadvertent omissions or human errors in compiling these experiments due to oversight.

Khalid Zaffar Masoodi: I dedicate this book to my Postdoc Mentor Prof. Zhou Wang, University of Pittsburgh, PA, USA for inculcating in me, the passion to do quality research through hard work and dedication. I also dedicate this work to my brother Er. Zahid Omer Masoodi for standing by me through thick and thin and as a strong support, to my daughter Afza Khalid Masoodi and my niece Liyana Masoodi as she wants to be a scientist like me and the Scientific Society who continue to inspire and support us.

Sameena Maqbool Lone: I dedicate this book to my father Mr. Mohammad Maqbool Lone, my mother Rafeeqa Maqbool Lone, my sisters Maryam M. Lone and Razia M. Lone, and my mentor Dr. Khursheed Hussain.

Rovidha Saba Rasool: I dedicate this book to my father M. Yousuf Mir, mother Sitara Jabeen, and niece Khadija.

Author contribution
All authors have contributed equally.

Introduction to molecular biology techniques

1

Molecular biology is a branch of science that deals with the study of molecules responsible for life. These molecules can be micro- as well as macromolecules. The three ideas that collectively can be said to form the science of molecular biology are Mechanism, Cell Theory, and Evolution. These ideas may help to further investigate the world of molecular biology. The subject mainly deals with the study of nucleic acids (DNA and RNA), proteins, and certain enzymes. With the combination of physics, chemistry, biochemistry, and genetics, this field of science moves down to the lowest answerable level of explanation. An important tenet of molecular biology is the way it explains the mechanisms of various biological phenomena.

In recent years, studies and investigations have been progressing at a fascinating rate in the field of biotechnology, molecular biology, and genetics. The credit for this progress goes to the development of new laboratory techniques that have been assisting in precise fractionation and analysis of macromolecules like nucleic acids and proteins, as well as the construction of recombinant DNA molecules that can then be cloned and expressed according to the requirements. Novel techniques have been regularly introduced and older techniques have been highly improved with the passage of time. The purpose of this book therefore is to bring the readers closer to the advanced analytical and preparative techniques that are considered to be used by the researchers, students, and workers more frequently in the field of biotechnology and molecular biology.

While choosing the techniques and methods to be incorporated in this book, we evidently had to be very selective and precise, and for the sake of comprehensiveness, the knowledge of certain basic biochemical techniques and terminology has been assumed. It has been seen that while performing the experiments students usually lag behind in knowing the actual background like the principle and applications of the method performed. Hence, while selecting these chapters we have taken into consideration all these aspects and observations.

In this manual, every chapter has been written with an aim to explain both the theory and relevant practical details along with the potentials and constraints for a given technique. The authors regularly use these techniques in their own laboratories of molecular biology, thus having the proper knowledge to provide step-by-step protocols and practical notes for each of the chapters. The detailed explanation of importance of each step, along with the importance of the chemicals and reagents used in the technique, is the distinguishing feature of this manual. The knowledge provided in the chapters can help the researchers to modify the techniques and adapt

Advanced Methods in Molecular Biology and Biotechnology. https://doi.org/10.1016/B978-0-12-824449-4.00001-3

them to different systems. This practical manual also provides a thorough concept of novel and economically efficient techniques used in the science of molecular biology, which can also be applied in many different fields.

In writing this practical manual, we have tried to tone it for students at a level that would be understandable by undergraduates though MSc and up to PhD students as well in the universities across India and the world as these are the stages at which we have experience of conducting these experiments.

Safety measures

To conduct the experiments, numerous hazards are associated. The chemicals like ethidium bromide (EtBr), acrylamide and phenols are highly toxic and even carcinogenic, which can result in permanent damage to one's body. So, care and precautions need to be taken to avoid these hazards. Appropriate care should be taken while working in UV radiations as it may cause severe damage to the eyes. The chemicals should be stored in their designated places and should be well labeled. The area or the laboratory should be kept clean. Lab coats, masks, gloves, and goggles should be worn while working in the laboratory.

Preparation of the solutions

(A) Calculation of the solutions—Molar, % and "×" (no. of times) solutions

- A 1-M solution contains 1-g molecular weight of the solute dissolved in the solvent to make 1 L of the final solution. For example, 100 mL of 5 M solution of NaCl contains 58.456 (MW of NaCl) g/mol × 5 mol/L × 0.1 L = 29.29 g of NaCl.
- Percentage (*w/v*) of the solution denotes the weight (g) of the solute in 100 mL of the final solution. For example, 0.7% solution of agarose in TBE buffer, will contain 0.7 g of agarose in the final volume of 100 mL TBE buffer.
- Enzyme buffers
- Many enzyme buffers need to be prepared as concentrated solutions, e.g., 5× (five times the concentration of the working solution), which are later diluted so that the final concentration of the buffer in the reaction is 1× (or as required). For example, 2.5 μL of a 10× buffer, other required components and water are mixed to make restriction digestion in 25 μL.

(B) Preparation of working solutions from concentrated stock solutions

Preparation of a stock solution for buffers is useful to avoid preparation of buffers every time one conducts an experiment. For example, 100 mL of TE buffer (10 mM Tris, 1 mM EDTA) will be made by combining 1 mL of 1-M Tris solution, 0.2 mL of 0.5 M EDTA, and 98.8-mL sterile water. For the calculations of the required stock solutions, following formula can be used:

$$C_i \times V_i = C_f \times V_f$$

where C_i=initial concentration or concentration of stock solution; V_i=initial volume, or amount of stock solution needed; C_f=final concentration, or concentration of desired solution; and V_f=final volume, or volume of desired solution.

(C) Steps for preparation of the solutions

- The laboratory manual and bottle labels should always be referred for the important instructions; and for any specific precautions specifically for preparation of the solutions.
- Required amount of chemicals should be weighed using an analytical balance if the amount is less than 0.1 g.
- The chemicals should be kept in appropriately sized beakers having a stir bar. Water should be added lesser than required and all the solutions should be prepared with double distilled water.
- The chemical when dissolved should be transferred to a graduated cylinder and required amount of distilled water should be added to make up the final volume.
- Solutions need to be autoclaved at 121°C for 20 min, whenever necessary. However, there are some solutions that cannot be autoclaved, e.g., SDS. Such solutions should be filter sterilized. Media to be used for bacterial cultures must be autoclaved the same day of preparation, preferably within an hour or two.
- The prepared solutions should be checked for contamination by holding the bottles containing the solution at eye level and giving it a gentle swirl before storing at room temperature. Solid media containing agar prepared for bacterial cultures, when needed, can be melted in a microwave, and if required additional components, e.g., antibiotics, can be added according to the need.

Concentrated solutions, e.g., 1-M Tris-HCl pH=8.0, 5M NaCl, can be used to make working stocks by adding autoclaved double-distilled water to the appropriate amount of the concentrated solution.

(D) Glassware and plastic ware

Glassware and plasticware should always be scrupulously clean. The contaminated glassware may result in the inhibition of the reactions and can also be responsible for the degradation of the nucleic acids. Glassware should be rinsed with distilled water before autoclaving at 150°C for 1 h. Micropipettes and microfuges should also be autoclaved. Glassware and solutions used for the experiments with RNA should be treated with diethyl-pyrocarbonate to inhibit RNases, which otherwise can be resistant to autoclaving. Plasticware such as pipette tips and culture tubes are often supplied sterile. The tubes that are being used should never come in contact with any other chemicals used in the experiments.

Working with the nucleic acids
Storage

There are certain properties of reagents and conditions that need to be considered in processing and storage of RNA and DNA. Free radicals that are formed from chemical breakdown and radiation can cause phosphodiester breakage in the nucleic acids. In addition, UV light at 260 nm can cause a variety of lesions, including thymine dimers and cross-linking, which can result in the loss of biological activity. Ethidium bromide is responsible for the photooxidation of DNA with visible light and molecular oxygen and these oxidation products can also result in phosphodiester breakage. Direct contact between the fingers and nucleic acids should always be avoided as nucleases are present on the human skin. Most DNases are not very stable; however, many RNases are very stable and can adsorb onto glass or plastic and remain active. For long-term storage of DNA, it is best to store at high salt concentration (> 1 M) in the presence of high concentration of EDTA (> 10 mM) at pH 8.5. Storage of DNA in buoyant CsCl with EtBr in the dark at 5 EC is excellent. There is about one phosphodiester break per 200 kb of DNA per year. Storage of λ DNA in the phage is better than storing the pure DNA.

Purification

For the removal of proteins:

- Treatment with proteolytic enzyme can be given, e.g., proteinase K. Silica-based column may be used.
- Extraction of DNA with phenol or phenol:chloroform and finally with chloroform can be done. Phenol is able to denature the proteins and the final extraction with chloroform removes traces of phenol.

Quantitation

- **Spectrophotometric**. A simple method of DNA quantitation is to read the absorbance at 260 nm where an OD of 1 in a 1-cm path length = 50 μg/mL for double-stranded DNA, 40 μg/mL for single-stranded DNA and RNA, and 20–33 μg/mL for oligonucleotides. An absorbance ratio of 260 and 280 nm gives an estimate of the purity of the solution. The method, though simple and efficient, is not useful for the quantities less than 1 μg/mL.
- **Ethidium bromide fluorescence**. The florescence emitted by EtBr in a solution is directly correlated with the amount of the DNA present in the solution dilutions of an unknown DNA in the presence of 2 μg/mL EtBr can be compared to dilutions of a known amount of a standard DNA solutions spotted in an electrophoresis gel.

Concentration

Nucleic acid solutions are concentrated with ethanol. The step-by-step methodology involved in the procedure is as follows:

1. Firstly, the volume of the DNA is measured and the concentration of monovalent cation is adjusted. The final concentration of ammonium acetate ($C_2H_7NO_2$), sodium acetate ($C_2H_3NaO_2$), sodium chloride (NaCl), and lithium chloride (LiCl) should be 2–2.5, 0.3, 0.2, and 0.8 M, respectively.
2. DNA fragments and oligonucleotides are precipitated with the help of the addition of $MgCl_2$ to the final concentration. The addition of monovalent cations is followed by the addition of 2–2.5 volumes of ethanol. The mixture is stored on ice at $-20°C$ for 20 min to 1 h.
3. The DNA can be recovered by centrifuging for 10 min at room temperature. The supernatant thus obtained is carefully separated, making sure that the DNA pellet is not discarded (often the pellet is not visible until it is dry).
4. The pellet is then washed with 0.5–1.0 mL of 70% ethanol to remove the salts, centrifuged again and supernatant decanted, and the pellet is dried. $C_2H_7NO_2$ is highly soluble in ethanol, thus can be efficiently removed with 70% wash.
5. Isopropanol is also used to precipitate DNA but it tends to coprecipitate salts and is harder to evaporate since it is less volatile. However, the requirement of isopropanol is less than ethanol for the precipitation of DNA.

Restriction enzymes

Restriction and DNA-modifying enzymes are stored at $-20°C$, typically in 50% glycerol. The enzymes can be stored in the insulated cooler as that would help to keep them at 20°C for some time. The tubes containing the enzymes should never be allowed to reach the room temperature. A new and sterilized pipette tip is used every time while using a restriction enzyme. In addition, the volume of the enzyme should be less than 1/10 of the final volume of the reaction mixture. Gloves should be worn while handling the enzymes, as human skin contains nucleases.

Sterile techniques

- All the media that is prepared should be autoclaved immediately in order to remove all kinds of contamination present in the media or in the glassware. The containers used for the media should be checked before pouring media into them by swirling them and checking for the cloudy substances at the center. A small amount of broth should be grown when the cells are needed to grow overnight. A small amount of contamination is not always apparent until the media is incubated at 37°C.

- The inoculating loops and tips of media bottles should be flame sterilized before and after pipetting. Always use fresh, sterilized pipettes while pipetting the media.
- The cultures that need to be grown overnight should always be grown on a fresh plate or previously tested glycerol stock that was grown from a single colony. For the preparation of an overnight culture grown on the glycerol stock, an individually wrapped 1 mL pipette and a culture tube of media are taken to the −80°C freezer. The cap should be removed quickly from the freezer containing the glycerol stock, a small amount of ice is scraped from the culture surface, the cap replaced, and the pipette placed in the culture tube. Sufficient numbers of bacteria are present in the ice in order for the culture to grow to saturation point in 16 h. Glycerol should never be thawed.

References

Crick, F. H. (1958, January). On protein synthesis. *Symposia of the Society for Experimental Biology, 12*, 138–163. 8.

Davis, R. W., & Botstein, D. (1980). *Advanced bacterial genetics: A manual for genetic engineering (no. 589.9 D3).*

Fry, M. (2016). *Landmark experiments in molecular biology.* Academic Press.

Slater, R. J. (1986). *Experiments in molecular biology* (pp. 121–129). Totowa, NJ: Humana Press.

Sablok, G., Budak, H., & Ralph, P. J. (2018). *Brachypodium genomics.* Springer.

Agarose gel electrophoresis

2

Definition

Electrophoresis is a method of separation of charged molecules under the influence of an electric field. It is an easy and most popular way of separating and analyzing DNA and other macromolecules with very high molecular weight. When charged molecules like DNA, RNA, proteins, etc. are kept in an electric field they migrate depending on their net charge, magnitude of electric field, shape, size in terms of their molecular weight, and media temperature. Since agarose gels are more porous than polyacrylamide gels due to comparatively larger pore size and DNA molecules being macromolecular in nature with uniform negative charge, they can easily be separated via agarose gel electrophoresis. DNA fragments get separated depending on their molecular weight, so that small fragments move at a faster pace in the matrix compared to large-sized molecules.

Rationale

The famous double helical structure of DNA reveals the negatively charged phosphate bond of DNA (or RNA) molecule. So, with the application of an electric current across the gel, DNA molecules starts migrating toward the opposite charged electrode, i.e., positive electrode (anode) through the matrix pores. A common question arises in young researchers that larger DNA fragments have more charge while small DNA fragments have less charge, so larger DNA fragments should be pulled by more force than the smaller DNA molecule, which is otherwise. Since both large and small DNA molecules have equal charge-to-mass ratio both are pulled by the same force toward the positive electrode. However, since it is easy for smaller DNA molecules to traverse through the gel matrix than the larger molecule, the smaller DNA molecules tend to move faster than the larger DNA molecule. Agarose gel electrophoresis involving larger DNA molecules require lower concentration of agarose gel and vice versa facilitating the rapid migration of small fragments toward the anode. The concentration of agarose can be varied and thus the pore size of the gel can be determined. Determination of the location of DNA in the gel can be done by staining

Advanced Methods in Molecular Biology and Biotechnology. https://doi.org/10.1016/B978-0-12-824449-4.00002-5

it with a very low concentration of intercalating fluorescent ethidium bromide dye under short-wave UV light, which highlights the fragments in the matrix.

It has been observed that the rate of migration of DNA through agarose gel depends on the following parameters:

- Molecular weight of DNA fragments—Smaller DNA molecules move faster than the larger ones.
- Agarose gel concentration forming matrix—Higher gel concentration has more resolving power than lower concentration gel. For larger DNA fragments, low concentration gel is used and vice versa.
- DNA conformation—One type of plasmid DNA gives multiple bands in gel because of different conformations of plasmid DNA. Circular plasmid DNA moves slowly while supercoiled plasmid DNA moves faster.
- Current applied—Higher current means more speed of migration. However, higher currents tend to increase the temperature of gel, which melts the gel.
- Ethidium bromide dye—EtBr has a positive charge and moves toward negative electrode; this makes the bands toward positive electrodes fainter than the bands toward the negative electrodes. Throughout the electrophoresis, EtBr travels continuously in the direction opposite to the direction of migration of DNA.
- Chemical composition of electrophoresis buffer-Maintenance of pH as well as conductance is determined by electrophoresis buffer. Usually, the buffer used for making the gel is used in electrophoretic tank for running the gel as well.

Before you begin

- Preparation of the gel bed.
- Preparation of DNA samples for running in the gel (DNA extraction, plasmid DNA extraction, restriction-digested DNA, etc.)

Key resources table

Reagent or resource	Source	Identifier
Agarose	Sigma-Aldrich	MFCD00081294
TAE buffer	Sigma-Aldrich	MFCD00236357
• Loading dye (6×)—10 mL		
• 0.25% Bromophenol Blue	Sigma-Aldrich	MFCD00013793
• 0.25% Xylene Cyanol FF	Sigma-Aldrich	MFCD00019481
• 30% Glycerol/Sucrose	Sigma-Aldrich	MFCD00004722
• EtBr	Sigma-Aldrich	MFCD00011724
DNA molecular weight ladder	Thermo Fisher scientific	12,352,019

Materials and equipments
Materials and equipments required
- Electrophoresis apparatus
- Electrophoretic tank
- Casting tray
- Comb
- Micropipettes
- Conical flasks for gel preparation
- Spectrophotometer
- Spatula

Reagents required
- 0.7% Agarose (0.7 g in 100 mL)
- Tris-acetate-EDTA (50×) TAE Buffer **(Stock Solution** = *50× and* **Working Solution** = *1×)*
- Tris-HCL (pH = 8.0)@242.2 g/L—Provides buffer conditions and maintains the pH of the system
- Acetic acid@60.5 mL/L—Helps in the conduction of electricity through the buffer
- EDTA@18.612 g/L—Acts as a chelating agent which chelates metal ions (Mg^{2+} in this case) which renders DNases inactive as Mg^{2+} acts as cofactors for DNases.
- EtBr **(Stock Solution** = *10 mg/mL and* **Working Solution** = *0.5 µg/mL)*—Helps in visualization of DNA as it has a planar structure like nitrogenous DNA bases. It intercalates between the bases and due to process of fluorescence enhancement DNA enhances the fluorescence of EtBr, which makes it fluoresce more than the surrounding gel.
- Loading dye (6×)—10 mL
- 0.25% Bromophenol Blue—Tracking dye—Helps visualize how much distance the gel has run.
- 0.25% Xylene Cyanol FF—Tracking dye—Helps visualize how much distance the gel has run.
- 30% Glycerol/Sucrose—Makes DNA heavier by increasing density and makes DNA settle in the well.
- DNA Molecular Weight Ladder

Step-by-step method details
Timing: 1–1.5 h (depending on the size of the sample)
- To prepare a 0.7% gel, add 0.7% agarose powder to the required volume of already prepared TAE buffer solution.
- Boil the solution at 100°C in order to dissolve the powder fully.

- Then, cool down the solution to approximately 55°C or when it is touchable with bare hands.
- After cooling, add the required volume of ethidium bromide to it (0.5 μg/mL).
- Pour the solution into a casting tray.
- Insert the comb across the end of the casting tray so that 1 mm gap between the teeth and the surface of tray could be made to form wells and allow the gel to solidify.
- When the gel solidifies, gently remove the comb and submerge it into 1 × TAE buffer-filled electrophoretic tank.
- With the help of a micropipette, load the samples (DNA) very carefully into the wells. However, before loading, mix the samples well with 6 × loading dye that will give higher density to the DNA sample so that they stay in the well.
- Connect a power source to the electrophoresis apparatus in such a way that negative terminal should be at the end where sample has been loaded and apply an electric current.
- Run the electrophoresis at 60–100 V till the Bromophenol Blue band migrates to the other end of the gel. The molecules having charge in them enter the gel matrix through its walls.
- Bubble formation takes place on the two platinum electrodes as a result of the electrolysis of water, which indicates the proper flow of current through the tank and is an indicator that gel has started running.
- After approximately 10 min, separation of the colored dyes becomes visible.
- Molecules having a negative charge like (DNA) in the gel will migrate toward the anode, while the molecules with positive charge shall migrate toward the opposite charged electrode, i.e., cathode.
- Once the gel has run significantly as visible from the Bromophenol Blue (tracking dye), observe the DNA under short-wave UV light on a trasilluminator or a gel doc.
- When the electrophoresis is completed, turn off the button and remove the electric leads to disconnect the power supply.
- Note down the gel results as well as data and gel data (Fig. 1).

Expected outcomes

DNA bands of different molecular weights are visible in the gel.

Quantification and statistical analysis

Quantification of gel electrophoresis is done by referring to the initial concentration of the DNA ladder and then the amount of DNA in each band is noted (amount of DNA of the ladder will be already given by the manufacturer).

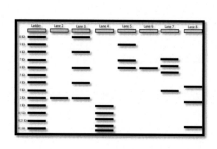 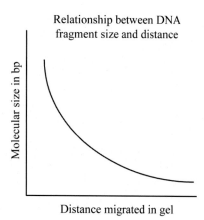

Gel data for agarose gel electrophoresis

FIG. 1

Gel data for agarose gel electrophoresis.

Advantages

Agarose gel electrophoresis is cheap and it can be processed easily. Recovery of samples is also more.

Limitations

Sometimes the gel melts when an electric current is passed and if it is made hard, gives poor results. The correct gel concentration for DNA samples needs to be used depending upon the expected size of DNA.

Optimization and troubleshooting

Problem

EtBr is known to have carcinogenic effects and has a heat labile nature as well. UV light is harmful to the eyes and the gel used in the technique can get spilled on the clothes.

Potential solution

Gloves should be used while handling the EtBr to avoid its hazardous effect. The solution should always be cooled down before adding EtBr with it. UV face shield gel doc system should be used to visualize the gels. An apron should be worn in the laboratory to avoid any kind of spillage on the clothes.

Safety considerations and standards

- Handle all the reagents involved with gentle care.
- Add EtBr only after cooling the solution, as it is heat labile.
- Mix the samples well with the loading dye before loading into wells.
- Avoid touching of EtBr with naked hands as it is highly carcinogenic.
- Note that most of the chemicals used in electrophoresis are highly toxic in nature.
- A UV shield should be worn when looking at the gel. Direct contact of eyes with UV can damage the eyes.

Alternative methods/procedures

- TBE-PAGE (polyacrylamide gel electrophoresis)
- Denaturing PAGE (polyacrylamide gel electrophoresis).
- SDS-PAGE (sodium dodecyl sulfate-polyacrylamide gel electrophoresis)

References

Aaij, C., & Borst, P. (1972). The gel electrophoresis of DNA. *Biochimica et Biophysica Acta, Nucleic Acids and Protein Synthesis, 269*(2), 192–200.

Das, S., & Dash, H. R. (2014). *Microbial biotechnology—A laboratory manual for bacterial systems*. Springer.

Kirkpatrick, F. H. (1990). Overview of agarose gel properties. In E. Lai, & B. W. Birren (Eds.), *Current communications in cell & molecular biology* (pp. 9–22). Plainview, NY: Cold Spring Harbor Laboratory Press.

Serwer, P. (1983). Agarose gels: Properties and use for electrophoresis. *Electrophoresis, 4*(6), 375–382.

Slater, R. J. (1986). Agarose gel electrophoresis of RNA. In *Experiments in molecular biology* (pp. 121–129). Totowa, NJ: Humana Press.

Smith, S. B., Aldridge, P. K., & Callis, J. B. (1989). Observation of individual DNA molecules undergoing gel electrophoresis. *Science, 243*(4888), 203–206.

Surzycki, S. (2003). Laboratory manual. In S. Surzycki (Ed.), *Human molecular biology laboratory manual*. Malden, MA: John Wiley & Sons. ISBN: 0-632-04676-7.

Wu, W., Zhang, H. H., Welsh, M. J., & Kaufman, P. B. (2016). *Gene biotechnology*. CRC Press.

SDS-polyacrylamide gel electrophoresis

Definition

Sodium dodecyl sulfate-polyacrylamide gel electrophoresis (SDS-PAGE) is a high-resolution technique commonly used for separating the mixture of proteins according to their molecular weight. SDS-PAGE involves the use of polyacrylamide as gel and sodium dodecyl sulfate (SDS) as ionic detergent to denature and linearize the proteins, hence, named so. This technique was developed by Ulrich K. Laemmli.

Rationale

- It has been seen that separation of some of the proteins does not occur due to identical charge:mass ratio. Therefore, such proteins are first treated with an ionic detergent called SDS before beginning and during the course of electrophoresis. It ensures that all the proteins in the sample reduce to their primary structures and carry a uniform negative charge; hence these properties cannot be manipulated for separation. Due to SDS treatment, the proteins denature and polypeptide chains get opened and extended leading to dissimilarity in their size in terms of differences in masses. On the basis of their mass but not the charge and shape, the molecules are separated.
- The SDS-treated proteins are denatured and have negative charge with identical charge-to-mass ratio. Proteins migrate toward the positive electrode through PAGE gel at alkaline pH. The small-sized polypeptides move at faster pace followed by the larger polypeptide units. Therefore, the intrinsic charge on proteins is masked in PAGE. Hence, the separation is based on molecular mass or size of polypeptides alone. β-Mercaptoethanol reduces the interpolypeptide disulfide bridges that are responsible for the protein folding, hence reducing them to their primary form, i.e., linear shape. The molecular weight of the separated protein can be analyzed by comparing the molecular weight and mobility of the standard protein.

Advanced Methods in Molecular Biology and Biotechnology. https://doi.org/10.1016/B978-0-12-824449-4.00003-7

Before you begin

- Preparation of the gel matrix.
- Preparation of samples for running in the gel.
- Preparation of proteins (RIPA protocol)

Key resources table

Reagent or resource	Source	Identifier
0.5 M Tris HCL	Sigma-Aldrich	MFCD00012590
10% SDS	Sigma-Aldrich	MFCD00036175
Acrylamide/bisacrylamide	Wiley/HIMEDIA	386284152/MB005
Ammonium per sulfate (APS)	HIMEDIA	MB003
Isopropanol	Sigma-Aldrich	MFCD00011674
SDS—Sample loading buffer (Lammeli Buffer 2×, pH = 6.8)	Sigma-Aldrich	S3401
SDS—Sample running buffer (Tris Gly, pH = 8.3)	Thermo Fisher scientific	LC2675
Coomassie staining solution	HIMEDIA	ML046
Destaining solution	Thermo Fisher scientific	P1304MP
Ladder	HIMEDIA	MBT051

Materials and equipment

Materials and equipment required:

- Gel casting apparatus (glass plates and spacers)
- Comb
- Clamps
- Micropipettes
- Conical flasks
- Spectrophotometer
- Spatula
- Distilled water
- Protein sample

Reagents required:

- **For 5-mL stacking gel:**
- DH_2O @ 2.975 mL
- 0.5 M Tris-HCL, pH = 8.0 @ 1.25 mL
- 10% SDS @ 0.05 mL
- Acrylamide/Bisacrylamide (30%/0.8% w/v) @ 0.67 mL

- 10% Ammonium persulfate (APS) @ 0.05 mL
- TEMED @ 5 μL
- **For 5-mL separating gel:**
- DH$_2$O @ 3.8 mL
- 1.5 M Tris-HCL (18.15 g Tris in 60 mL of distilled water and make the final volume to 100 mL (pH = 8.8) @ 2.6 mL—Maintains pH
- 10% SDS @ 0.1 mL—Denatures protein and provides uniform charge to proteins.
- Acrylamide/Bis-acrylamide (30%/0.8% w/v) @ 3.4 mL—Polymerizes and forms a matrix
- 10% Ammonium persulfate (APS) @ 100 μL—Provides free radicles for matrix formation; *prepare fresh*
- TEMED @ 10 μL—Acts as a catalyst for the initiation of polymerization.
- **Isopropanol**—Blocks O$_2$ from reacting with gel as O$_2$ can inhibit polymerization. The gel is placed between two glass plates so that O$_2$ does not come in contact with gel as oxygen hampers gel formation. This is also the reason for using isopropanol. In addition, it removes bubbles and gives a uniform shape to the upper gel border.
- **For Laemmli Procedure of SDS—Sample loading buffer (Laemmli buffer 2×, pH = 6.8)**
- 4% SDS—Denatures protein and provides uniform charge to proteins
- 10% 2-Mercaptoethanol—Reducing agent, breaks disulfide bridges, and converts tertiary structure of protein into primary structure.
- 20% Glycerol—Increases density of protein and makes it settle in the wells after loading.
- 0.004% Bromophenol Blue—Tracking dye that helps to track movement of proteins in gel so that they do not overrun.
- 0.125-M Tris HCL
- **SDS—Sample running buffer (Tris Gly, pH = 8.3)** *(***Stock Solution** = *10× and* **Working Solution** = *1×)—Freshly prepared*—Glycine ions and chloride ions have a big role in SDS-PAGE. Since there is a difference in pH between the stacking gel (pH 6.8) and resolving gel (pH 8.8), stacking gel removes bias between the proteins that enter the gel first and the proteins that enter the well last. Since well is ≥ 10 mm in height, some proteins enter the gel first and some enter the gel last. This causes a bias in resolution. To overcome this, the stacking gel is used. Proteins entering the gel are sandwiched between glycine ions and chloride ions. Glycine ions behave differently in stacking and resolving gel owing to the difference of pH of these two gels. At pH 6.8, the glycine and chloride ions have lesser negative charge and they sandwich the protein and make it into a flat disc so that when the proteins are about to enter the resolving gel all proteins enter the stacking gel together. Upon entering the resolving gel at pH 8.8, the chloride ions move fast and glycine ions move after them leaving the proteins behind to resolve according to the molecular weight.

- **Coomassie staining solution**: Fixes the proteins in the gel. If this step is not carried out then proteins will be washed in the solution and bands will not be visible
- Ethanol @ 150 mL—Keeps the gel in shape or else the gel will become bloated in water and swell. Also help precipitate the proteins in the gel for the dye to interact.
- Glacial acetic acid @ 50 mL—Acetic acid provides acidic medium and increases interaction of proteins with dye so that they are visible under white light.
- DDH$_2$O @ 300 mL
- CBB-R-250 @ 1 g—Stains proteins and makes them visible

Dissolve CBB-R-250 in EtOH first

- **Destaining solution:**
- Ethanol @ 1200 mL
- Glacial acetic acid @ 400 mL
- DDH$_2$O @ 2.4 L
- **Ladder**

Step-by-step method details
Timing: 3–4 h (depends on the plate size)

- **Casting the gel:**
- Put together the glass plates and the spacers in gel casting apparatus.
- Mix the components of stacking gel in the order given above.
- Put the resolving gel mixture carefully into the gel plate to a level 2 cm below the top of the shorter plate with the help of a micropipette.

Note: To prevent bubbling, a layer of isopropanol can be placed over the top of the resolving gel, to prevent oxygen from reacting with the gel, and to straighten the level of the gel by removing the bubbles by decreasing the surface tension.

Pause point: Leave resolving gel to stand for 20–30 min at the room temperature.

- Once the gel has polymerized, the isopropanol from the top of the resolving gel should be completely drained. Rinse with DH$_2$O for 5–7 times to remove any traces of isopropanol.
- Mix the stacking gel components.
- Pour the stacking gel (other than separating gel) solution into gel plates (on top of running gel), so that gel plates are completely filled.
- Then, insert the comb at the top of the stacking gel for polymerization.

Pause point: Allow gel to stand as such for 20–30 min at room temperature.

- **Preparing samples:**
- **Cell samples (use RIPA buffer protocol for protein extraction from cells)**
- Harvest 100 μL of cells at OD > 0.6.

- Pour the supernatant media off.
- Resuspend the cells in 20 μL of 2× sample buffer
- Incubate tubes in the boiling water for 5 min to denature protein content.
- Allow centrifugation at 12,000×g for 30 s.
- **Solution samples:**
- Pour a volume of protein solution (@ 5 μL) into the centrifuge tube.
- Load an equal volume of 2× sample buffer.
- Incubate the tubes that contain the samples, in boiling water for 5 min for denaturation purpose.
- Allow centrifugation at 12,000×g for 30 s.
- **Running the gel:**
- Remove the comb to prevent sample wells from deformation and assemble the cast gel.
- Carefully add freshly prepared 1x running buffer (pH = 8.3) to both chambers of the electrophoresis apparatus.
- With the help of a syringe, the wells are washed with the running buffer to remove any un-polymerized gel. This may interfere with the protein resolution. Cleaner wells give good results.
- With the help of a micropipette, pour the already prepared samples into the wells of the gel.
- Tightly connecting the electrodes with a power supply, run the gel at 100 V until the Bromophenol Blue dye front migrates into the running gel (~15 min) and increase to 200 V until the dye front moves completely to the bottom of the gel matrix (~45 min).
- **Staining and destaining the gel:**
- Remove spacers, stacking gel, and glass plates from the gel setup.
- Transfer the gel to the staining tray containing the gel staining dye.
- Add 20 mL of Coomassie Blue stain and stain for > 30 min with gentle shaking.
- Decant the stain.
- Carefully add ~5-mL destaining solution and destain for 1 min by shaking gently.
- Decant and discard the destaining solution.
- Add ~30 mL of fresh destain solution and destain with gentle shaking. Repeat this (pouring and decanting of the destaining solution) until the bands are clearly visible.
- Decant and discard the destaining solution.
- Wash quickly with DH$_2$O.
- With gentle shaking, add ~30 mL DH$_2$O for 5 min and rinse.
- Put a sheet of Whatmann filter paper and a piece of saran below and over the gel surface respectively. Allow the gel to dry on the gel dryer at 60°C for 1 h. Visualize the gel.

Expected outcomes

Protein bands of different molecular weights are visible in the gel (Fig. 2).

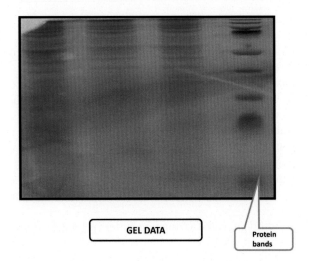

GEL DATA

Protein
bands

FIG. 2

SDS-PAGE gel depicting various bands of proteins (in dark blue color) of different molecular weights resolved. Higher molecular weight proteins move slowly and are near the well and low-molecular weight protein bands move fast and are near the lower end of the gel.

Quantification and statistical analysis

The migration of the proteins when compared to the set of standard proteins run on the same gel helps to calculate the size of the proteins For example, a worker using a 7.5% gel will choose the standards with higher molecular weights than a worker using a 15% gel, which can be better suitable for the analysis of small proteins. A plot of the $\log_{10}$ MW of the standard proteins against the distance that each protein migrated on the gel will give a straight line in the region where the gel has good resolving power. The sizes of unknown proteins can be determined by incorporating experimental values on a graph of standard proteins. Proteins whose molecular weights are beyond this range will not be well resolved on the gel.

Advantages

Polyacrylamide gel is chemically stable and a small quantity of sample may be required for this technique. In addition, the protein resolution is high while using this method.

Limitations

The technique is quite costly. In addition, the time taken for gel preparation is longer than the time taken in agarose gel preparation. Acrylamide and bisacrylamide are toxic in nature.

Optimization and troubleshooting
Problem

Bubbles are usually formed during pouring of the gel. The staining dyes have carcinogenic effects and the samples get denatured.

Potential solution

Gel should be poured from the sides and the layer of isopropanol should be placed on the top of the resolving gel. Gloves should always be worn before handling the staining dyes, and to avoid denaturation, the samples should always be kept on ice before loading.

Safety considerations and standards

- Ensure there is no leakage in the gel caster plate, before you add the resolving gel.
- Ensure no bubbles are left on the face of the stacking gel and in the wells. This will ensure better protein resolution.
- Do not forget to wear standard gloves while handling the TEMED and gel.
- Before loading keep all the samples on ice.
- Avoid heating the sample for more than 5 min.
- To avoid overstaining that could possibly delay results, do not stain the gel more than 5 min.
- Handle the gel carefully considering the delicate nature of the SDS-PAGE gel.

Alternative methods/procedures

- Gel filtration
- Affinity chromatography
- Isoelectric focusing

References

Dunbar, B. S. (1994). *Protein blotting: a practical approach (No. 547.96 PRO)*.

Geisberg, J. V., & Moqtaderi, Z. (2019). Protein binding to mRNA 3′ isoforms. *Current Protocols in Molecular Biology*, *128*(1), e101.

Gels, T., Schagger, H., & von Jagow, G. (1987). Tricine-sodium dodecyl sulfate-polyacrylamide gel electrophoresis for the separation of proteins in the range from 1 to 100 kDa. *Analytical Biochemistry*, *166*, 368–379.

Gorman, K., McGinnis, J., & Kay, B. (2018). Generating FN3-based affinity reagents through phage display. *Current Protocols in Chemical Biology*, *10*(2), e39.

Smith, B. J. (2009). Chemical cleavage of proteins at asparaginyl-glycyl peptide bonds. In *The protein protocols handbook* (pp. 899–903). Totowa, NJ: Humana Press.

Immunoelectrophoresis

4

Definition

Immunoelectrophoresis is a technique used for the characterization of antibodies. The term "immunoelectrophoresis" was coined by Grabar and Williams in 1953. The principle behind this technique is the electrophoresis of antigen for immunodiffusion with a poly-specific antiserum to form precipitin bands. The molecules possessing the charge move toward the appropriate electrode. The migration of the molecules depends on a number of factors such as net charge, size, pH of the buffer, and ionic strength of the buffer. Therefore, when an antigen is subjected to electrophoresis in an agarose gel, it gets separated according to the properties it possesses. The antigen thus resolved is subjected to immunodiffusion with antiserum added in the agarose gel. The antibodies diffuse laterally to meet diffusing antigens, and lattice formation and precipitation occur permitting determination of the nature of the antigens. The precipitin line indicates the presence of the antibody, specific to antigen.

Before you begin

- Preparation of gel bed.
- Preparation of electrophoresis buffer

Key resources table

Reagent or resource	Source	Identifier
Agarose	Sigma-Aldrich	MFCD00081294
TAE buffer	Sigma-Aldrich	MFCD00236357
• Loading dye (6×)—10 mL		
• 0.25% Bromophenol Blue	Sigma-Aldrich	MFCD00013793
• 0.25% Xylene Cyanol FF	Sigma-Aldrich	MFCD00019481
• 30% Glycerol/Sucrose	Sigma-Aldrich	MFCD00004722
• EtBr	Sigma-Aldrich	MFCD00011724
DNA molecular weight ladder	Thermo Fisher scientific	12352019
Test Antiserum	HIMEDIA	–
Antibody	HIMEDIA	

Advanced Methods in Molecular Biology and Biotechnology. https://doi.org/10.1016/B978-0-12-824449-4.00004-9

Materials and equipment

Materials and equipment required:

- Electrophoresis apparatus
- Electrophoretic tank
- Casting tray
- Comb
- Micropipettes
- Conical flasks for gel preparation
- Spectrophotometer
- Spatula

Reagents required:

- Agarose
- Tris-acetate-EDTA (50×) TAE buffer *(**Stock Solution** = 50× and **Working Solution** = 1×)*
- Tris-HCL (pH = 8.0)—Provides buffer conditions and maintains the pH of the system
- Acetic acid—Helps in the conduction of electricity through the buffer
- EDTA—Acts as a chelating agent which chelates metal ions
- Ethidium bromide (EtBr) *(**Stock Solution** = 10 mg/mL and **Working Solution** = 0.5 µg/mL)*—Helps in the visualization of macromolecule. Loading dye (6×)—10 mL
- 0.25% Bromophenol Blue—Tracking dye—Helps visualize how much distance the gel has run
- 0.25% Xylene Cyanol FF—Tracking dye—Helps visualize how much distance the gel has run
- Test antiserum A
- Test antiserum B
- Antibody to whole serum
- Antigen

Step-by-step method details

Timing: 2 days (Day 1: 3 h 30 min: electrophoresis and immunodiffusion. Day 2: 30 min: observation and interpretation)

- 10 mL of 1.5% agarose is prepared.
- The side of the glass plate to be placed toward negative electrode during electrophoresis is marked.
- The solution is cooled to 55–60°C and 6 mL/plate is poured onto a grease-free glass plate placed on a horizontal surface. The gel is allowed to set for 30 min.

- The glass plate is placed on the template provided. A well is punched with the help of the gel puncher corresponding to the markings on the template at the negative end. Use gentle suction to avoid forming rugged wells.
- Two troughs are cut with the help of the gel cutter, without removing the gel from troughs. Then 10 μL of the antigen is added to the well and the glass plate is placed in the electrophoresis tank such that the antigen well is at the cathode/negative electrode.
- 1× Electrophoresis buffer is poured into the electrophoresis tank such that it just covers the gel.
- The antigen is subjected to electrophoresis at 80–120 V and 60–70 mA, until the blue dye travels 3–4 cm from the well.

Note: Do not electrophorese beyond 3 h, as it is likely to generate heat.

- After electrophoresis, the gel is removed from both the troughs and the plate is kept at room temperature for 15 min. Then 80 μL of antiserum A is added in one of the trough and antiserum B in the other.
- The glass plate is placed in a moist chamber and incubated overnight at 37°C. Precipitin lines between antiserum troughs and the antigen well are observed.

Expected outcomes

The formation of precipitin line signifies the presence of antibody specific to the antigen.

- The formation of elliptical precipitin arcs indicates antigen-antibody interaction.
- The absence of precipitate suggests no reaction.
- Different antigens (proteins) can be identified on the basis of shape, position, and intensity of the precipitation lines.

Advantages

Immunoelectrophoresis is a powerful analytical technique, which combines electrophoresis for the separation of antigens with immunodiffusion against an antiserum. The prime advantage of immunoelectrophoresis is that it enables the identification of a number of antigens in serum.

Limitations

Immunoelectrophoresis is slower, less sensitive, and more difficult to interpret than immunofixation electrophoresis. The presence of small M-proteins may be obscured by the rapidly migrating immunoglobulins present in high concentrations, thus leading to failure in detection of some small monoclonal M-proteins by IEP. The

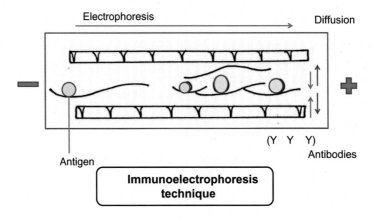

FIG. 3

Immunoelectrophoresis technique.

unavailability of specific antibodies limits the use of immunoelectrophoresis in food analysis (Fig. 3).

Optimization and troubleshooting
Problem
Usually either no or improper (usually blue) precipitin ring is observed.

Potential solution
Samples must be loaded directly into the well and troughs avoiding spilling to the sides. Avoid drying of the gel by keeping enough moist cotton in the moist chamber. The antigen should travel to at least three-fourth of the gel.

Safety considerations and standards
- Wear gloves and safety goggles at all times as good laboratory practice.
- Be careful with equipment that use heat and mix of reagents together.
- Wash hands with soap and water after using biological reagents and materials.
- Always wear safety goggles and use hot gloves while heating agarose as it may splatter causing burns.
- Ensure that the chamber is in level and the work surface is dry before turning the power supply on.
- Add EtBr only after cooling the solution, as it is heat labile.
- Mix the samples well with the loading dye before loading into wells.

- Note that most of the chemicals used in electrophoresis are highly toxic in nature.
- UV shield should be worn when looking at the gel. Direct contact of eyes with UV can damage the eyes.

References

Actor, J. K. (2019). *Introductory immunology, 2nd: Basic concepts for interdisciplinary applications.* Academic Press.

Martin, F., & Uroz, S. (Eds.). (2016). *Microbial environmental genomics (MEG)* Humana Press.

Parija, S. C. (2014). *Textbook of microbiology & immunology-E-book.* Elsevier Health Sciences.

Sastry, A. S., & Bhat, S. (2018). *Essentials of medical microbiology.* Jaypee Brothers, Medical Publishers Pvt. Limited.

Plasmid DNA isolation using alkaline lysis method from *E. coli* DH5α cells

5

Definition and rationale

Plasmids are small, double-stranded, self-replicating, covalently bonded DNA molecules that occur naturally in many prokaryotic and eukaryotic organisms like bacteria, fungi, cyanobacteria, etc. They range in size from 1 kb to over 300 kb. There are three forms of plasmids: linear, supercoiled, and relaxed. The alkaline lysis method for isolating plasmid DNA takes advantage of the physical difference between linear, closed, and supercoiled plasmid DNA and genomic DNA. In this method, the bacterial cells like that of *E. coli* are lysed using EDTA and an SDS detergent. This weakens the bacterial cell wall and also inactivates the enzymes that digest the DNA. The bacterial proteins are denatured by SDS treatment. This helps to rupture the cell wall by hyperlytic osmosis releasing the DNA (chromosomal and plasmid), proteins, and other contents. After adding sodium hydroxide and increasing the pH, the chromosomal and plasmid DNAs denature. With the reduction of the pH of the solution, the plasmid DNA renatures faster than genomic DNA because of its small size and highly supercoiled conformation. The chromosomal DNA and bacterial proteins get precipitated and are removed by centrifugation. Plasmid DNA gets precipitated by isopropanol/ethanol addition. Subsequently, other salt contaminants are removed using 70% ethanol.

Before you begin

- Preparation of different kinds of lysis solutions.
- Preparation of bacterial culture.
- Preparation of gel bed.

Advanced Methods in Molecular Biology and Biotechnology. https://doi.org/10.1016/B978-0-12-824449-4.00005-0

Key resources table

Reagent or resource	Source	Identifier
LB broth	HIMEDIA	M1245
M Tris-HCL stock solution	HIMEDIA	MB030
EDTA	Sigma-Aldrich	MFCD00070672
SDS	Sigma-Aldrich	MFCD00036175
Potassium acetate	Sigma-Aldrich	MFCD00012458
Glacial acetic acid	Sigma-Aldrich	MFCD00036152
Isopropanol	Sigma-Aldrich	MFCD00011674
Tris-EDTA	Sigma-Aldrich	MFCD00236359

Materials and equipment

Materials and equipment required:

- Centrifuge
- 1.5-mL microfuge tubes
- Micropipettes
- Agarose electrophoresis equipment and reagents
- Shaking incubator

Reagents required:

- **LB broth:**
- 200 mM NaCl
- 1% Tryptone
- 0.5% Yeast extract
- **Alkaline lysis Solution I** *(resuspension solution—store at 40°C):*
- 500-mM Glucose stock solution—provides isotonic medium
- 25-mM Tris-HCL stock solution (pH = 8.0)—maintains pH
- 10-mM EDTA—Acts as a chelating agent that chelates metal ions (Mg^{2+} in this case) and renders DNases inactive and prevents them from degradation.
- **Alkaline lysis Solution II** *(lysis solution—store at room temperature):*
- 0.2 N NaOH *(Stock Solution = 0.4 N)*—Provides alkaline environment for lyses and denaturing the DNA
- 1% SDS *(Stock Solution = 2%)*—Helps in lysis of cells.
- **Alkaline lysis Solution III** *(neutralizing solution—store at room temperature):*
- 3-M potassium acetate (pH = 6.0)—Furnishes K^+ ions which bind to the phosphate backbone of DNA and reduces repulsive negative force between two DNA strands and helps them to renature easily.
- 11.5% glacial acetic acid—Brings down pH at which DNA starts to renature.

- **Isopropanol (equal volume)/absolute ethanol (double volume).** It causes DNA precipitation and helps in separating out DNA from the solution by bringing down the dielectric constant of the solution.
- **70% Ethanol**—Used for washing DNA. It allows salts to dissolve and leaves DNA insoluble
- **TE: Tris-EDTA**
- 1 mM EDTA—Chelates metal ions and helps in protecting DNA from nucleases.
- 10 mM Tris-HCL (pH = 8.0)—Maintains the pH of the solution for longer stability of DNA and storage.

Step-by-step method details

- Take a loopful of culture of *E. coli* harboring a plasmid (pEGFPC1 in this case) and streak onto Luria Bertani medium agar plate with kanamycin (working concentration of 25 μg/mL). This plasmid contains a kanamycin resistance gene. Use an antibiotic depending on the antibiotic marker gene present.
- Incubate it overnight at 37°C.
- On the following day, pick up a single colony from the LB agar plate and inoculate it in 10 mL of LB broth containing 25 μg/mL of kanamycin.
- Incubate the test tube overnight at 37°C under constant shaking in a shaker incubator.
- Take 1.5 mL of the incubated culture (broth) into a micro-centrifuge tube and centrifuge the cells at 5000 *g* for 5 min. Discard the supernatant completely.
- Completely resuspend the cell pellet in 150 μL of ice cold resuspension solution (Solution I) and mix or vortex to get a uniform suspension without any clumps being visible; the pellet is suspended in Solution I.
- For cell lysis add 200 μL of Lysis solution (Solution II—freshly made) to lyse the bacterial cells. Seal the tube tightly and invert the tube many times until the solution becomes viscous.
- Add 300 μL of neutralizing solution (Solution III) and invert the tube many times to mix the content thoroughly and then keep the tube on ice for 2 min.

Critical: Do not exceed 2 min as this is the crucial step for plasmid DNA isolation. If the tube is kept longer than 2 min, the genomic DNA will also renature and contaminate our plasmid.

- Centrifuge the sample at 12,000 rpm for 15 min.
- Transfer the supernatant containing plasmid DNA to a 1.5 mL microfuge tube carefully without disturbing the white pellet.
- Carefully add equal volume of isopropanol to the tube and mix the content by inverting the tube for 5 min at room temperature to precipitate the plasmid DNA.

- Centrifuge the tube for 15 min at 12,000 rpm at 24°C.
- White/watery pellet of plasmid DNA become visible at the bottom of the tube.
- Thereafter remove the supernatant and wash pellet in 500 µL of 70% ethanol.
- Discard the supernatant and dry the tube containing pellet at 37°C so that no trace of isopropanol is left.
- Dissolve the dry pellet in 80 µL TE—for long-term storage at − 20°C.
- Thereafter, the sample is subjected to electrophoresis (agarose gel) to check the plasmids.

Expected outcomes

Multiple bands for each plasmid group are observed due to different confirmations (Fig. 4)

Advantages

This technique is fast and reliable.

Limitations

It can be costly as every time freshly made lysis solutions are used.

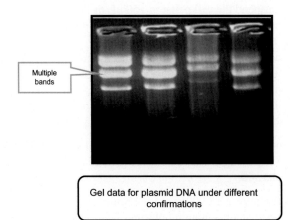

Multiple bands

Gel data for plasmid DNA under different confirmations

FIG. 4

Gel data for plasmid DNA under different confirmations.

Optimization and troubleshooting
Problem
Genomic DNA gets sheared and there is poor precipitation.

Potential solution
After adding the solution, the contents should not be mixed vigorously or vortexed. A chilled neutralizing solution should be used and incubation should be done on ice to avoid poor precipitation.

Safety considerations and standards
- Ensure that the bacterial cell pellet is uniformly dispersed in resuspension buffer before it is lysed.
- After adding Solution II, do not mix the contents by vigorously shaking or vortexing as it could lead to shearing of genomic DNA and will contaminate the plasmid DNA.
- Always use freshly made lysis solution.
- Do not forget to use chilled neutralizing solution and incubation on ice for 2 min to enhance better precipitation.
- While transferring the supernatant ensure that the white pellet does not come along with the transferring supernatant.
- Close the tube tightly before mixing the content within by inverting it and put on ice for the mentioned time period to increase plasmid yield.
- Do not over dry the pellet as overdried pellet is difficult to dissolve.

Alternative methods/procedures
- Plasmid DNA isolation using amino-silica-coated magnetic nanoparticles (ASMNPs).
- Plasmid DNA isolation using sodium lauryl sulfate and NaCl.
- Plasmid DNA isolation using Plasmid Mini Kit.
- NucleoSpin Plasmid Kit.
- Exogenous plasmid isolation (Kyselková et al., 2016).
- Transposon-aided capture of plasmids (TRACA).
- Multiple displacement amplification.

References

Bimboim, H. C., & Doly, J. (1979). A rapid alkaline extraction procedure for screening recombinant plasmid DNA. *Nucleic Acids Research, 7*(6), 1513–1523.

Birnboim, H. C. (1983). A rapid alkaline extraction method for the isolation of plasmid DNA. In *Vol. 100. Methods in enzymology* (pp. 243–255). Academic Press.

He, F. (2011). Plasmid DNA extraction from E. coli using alkaline lysis method. *Biology Protocol, 30*, 1–3.

Kyselková, M., Chrudimsky, T., Husník, F., Chronáková, A., Heuer, H., Smalla, K., et al. (2016). Characterization of tet(Y)-carrying LowGC plasmids exogenously captured from cow manure at a conventional dairy farm. *FEMS Microbiology Ecology, 92*, fiw075. https://doi.org/10.1093/femsec/fiw07.

Restriction digestion of plasmid DNA

6

Definition and rationale

Restriction enzymes are commonly called "Molecular Scissors." They recognize and cleave the specific DNA sequences called restriction sites. Restriction digestion is used as a tool to carry out manipulation of genetic material. It takes advantage of these naturally occurring enzymes. Restriction digestion is heavily used in modification and restriction mechanism of DNA. It is useful in gene cloning, DNA amplification, restriction cloning, etc.

Before you begin

- Preparation of DNA samples.
- Recognition of specific restriction enzyme.

Key resources table

Reagent or resource	Source	Identifier
Biological samples		
Liquid plasmid DNA aliquot		
Restriction enzyme	HIMEDIA	–
Assay buffer for restriction enzyme	HIMEDIA	–
Bromophenol blue gel loading dye	Sigma-Aldrich	MFCD00013793
Agarose gel electrophoresis buffer	Sigma-Aldrich	MFCD00081294

Materials and equipment

Materials and equipment required:

- Electrophoresis equipment.
- Micropipettes and pipette tips.

Reagents required:

- Liquid plasmid DNA aliquot.
- Restriction enzyme (e.g., Pst I).
- Assay buffer for restriction enzyme.
- Bromophenol Blue gel loading dye.
- Agarose gel electrophoresis buffer.

Step-by-step method details
Timing: 2–3 h

- Thaw all reagents on ice.
- Take a 1.5 mL tube; in the following order, combine the following reagents sequentially:
- Water—Final volume to 20 μL
- Buffer—10x, 2 μL
- Plasmid (1 μg/μL concentration)—1 μL
- Restriction enzyme—1 Unit
- Mix the contents by tapping the tube and then spinning the contents for 15–20 s
- Incubate the tube at an appropriate temperature, usually 37°C for 1 h

The results are then analyzed via gel electrophoresis.

Expected outcomes

The *E. coli* plasmid pEGFP-C1 shows restriction sites of different restriction enzymes in the current experiment (Fig. 5).

Advantages

It is a simple and efficient technique of digestion of plasmid DNA

Limitations

The specificity of restriction enzymes.

Optimization and troubleshooting
Problem

Restriction enzymes get denatured and certain enzymes are unable to cleave.

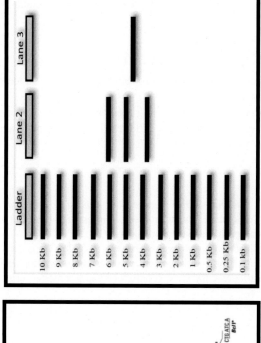

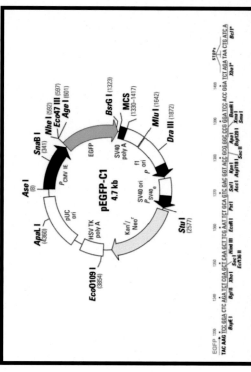

FIG. 5

(A) *E. coli* plasmid pEGFP-C1 shows restriction sites of different restriction enzymes. (B) Gel data.

Potential solution

Restriction enzymes should be placed on ice bucket after removing them from $-20°C$.

Safety considerations and standards

- Incubate for the specified time period at room temperature or on ice so as to obtain good results.
- Add a restriction enzyme only after adding the appropriate restriction digestion buffer as it creates perfect conditions for the restriction enzyme to act.

References

Schagat, T. (2007). *Rapid DNA digestion using Promega restriction enzymes.* Promega Corporation. www.promega.com/resources/articles/pubhub/enotes/rapid-dna-digestion-using-promegarestriction-enzymes/.

Tritle, D. (2006). *Activity of Promega restriction enzymes in GoTaq Green and PCR Master Mixes.* Promega Corporation. www.promega.com/resources/articles/pubhub/enotes/activity-of-promega-restriction-enzymesin-gotaq-green-and-pcr-master-mixes/.

Genomic DNA extraction from the plant leaves using the CTAB method

Definition

DNA is found in all living organisms such as viruses, plants, animals, etc. It is a high molecular weight macromolecule that serves as a blueprint of life. The entire genetic message that controls the chemistry of every cell is programmed in the DNA. The extraction of DNA from plant tissues is a difficult task owing to the presence of a rigid cell wall that guards the plant cells. The extraction of DNA essentially requires any procedure that breaks the cell wall and membranes allowing access to nuclear material, without its degradation. The CTAB method is a forerunner choice that can be applied to isolate genomic DNA from both fresh and dried leaves.

This method usually involves an initial grinding stage to break down the plant cell wall material to allow access to DNA. The ground tissues are then suspended in a CTAB buffer. In order to purify DNA, insoluble particulates are removed by centrifugation while mixing with chloroform:Isoamyl alcohol followed by centrifugation aids in the separation of soluble proteins and reducing foam. DNA is then precipitated from the aqueous phase and washed thoroughly to remove contaminating salts. The purified DNA is then resuspended and stored in a Tris-EDTA buffer. In order to check the quality of the extracted DNA, a sample is run on an agarose gel (0.7%), stained with EtBr and visualized under UV light.

Rationale

- A cationic detergent called Cetyl Trimethyl Ammonium Bromide (CTAB) is employed to cause lysis in the plant cells and dissolve its contents. This is followed by disruption of the cell membrane to release the DNA into the extraction buffer. In order to protect the obtained DNA from endogenous nuclease enzymes, a chelating agent EDTA is often included in the extraction buffer which chelates the magnesium ions, a necessary cofactor for nucleases. The incipient DNA extract contains a large amount of RNA, proteins, polysaccharides, polyphenols (tannins), and pigments with it, which are difficult to separate and may interfere with the DNA extraction.

Advanced Methods in Molecular Biology and Biotechnology. https://doi.org/10.1016/B978-0-12-824449-4.00007-4

- Separation of polysaccharides from DNA depends on their solubility in the CTAB buffer, which varies with the concentration of NaCl. At a higher concentration of NaCl, polysaccharides are insoluble, while at a lower concentration of NaCl, DNA is insoluble. Subsequently, by modifying the concentration of salt, polysaccharides and DNA can be differentially precipitated.
- Polyphenols, the compounds that occur naturally in plants, are also formed when plant tissues get damaged (Browning) and bind very efficiently to DNA. The polyphenol oxidase liberated upon the homogenization of plant tissues synthesizes polyphenols. To prevent the phenolic rings from binding up with the DNA molecules, polyvinyl pyrrolidone (PVP) is added.
- Most of the proteins are separated from the DNA extract using chloroform and isoamyl alcohol, which cause denaturation of proteins and precipitation.
- On the other hand, RNA contaminants are removed by treating the DNA extract with RNase.
- The DNA is then precipitated out and washed in organic solvents before redissolving in aqueous solution.

Before you begin

- DNA extraction from the plant leaves.
- Preparation of gel matrix.

Key resources table

Reagent or resource	Source	Identifier
CTAB @ 2.0g—CTAB (cationic detergent)	Sigma-Aldrich	MFCD00011772
Tris-HCL	Sigma-Aldrich	MFCD00012590
EDTA	Sigma-Aldrich	MFCD00070672
NaCl	Sigma-Aldrich	MFCD00003477
1% PVP-40	HIMEDIA	RM854
Isopropanol	Sigma-Aldrich	MFCD00011674
Tris EDTA buffer	Sigma-Aldrich	MFCD00236359

Materials and equipment

Materials and equipment required:

- Centrifuge
- 2 mL microfuge tubes
- Mortar and pestle

- Water bath (60–65°C)
- Sterile water
- Micropipettes
- Agarose electrophoresis equipment and reagents.
- Shaking incubator

Reagents required:
   2% CTAB buffer (pH = 5.0):

- CTAB @ 2.0 g—CTAB (cationic detergent) is used to lyse the plant cells and dissolve its contents and also undergoes complex formation with DNA and helps in its extraction.
- Tris–HCL (100 mM) @ 10 mL—maintains pH (pH = 8.0).
- EDTA (20 mM) @ 4 mL—EDTA chelates Mg^{2+} ions, a necessary cofactor for nucleases, and prevents degradation of DNA.
- NaCl (1.4 M) @ 28 mL—DNA and polysaccharides show differential solubility in CTAB depending on the concentration of NaCl. At a higher concentration of NaCl, polysaccharides are insoluble, while at a lower concentration of NaCl, DNA is insoluble. Subsequently, by modifying the salt concentration, polysaccharides and DNA can be differentially precipitated.
- H_2O @ 40 mL
- 1% PVP-40 @ 1 g—PVP prevents the phenolic rings from binding up with the DNA molecules.
- **Isopropanol (Equal Volume)/Absolute ethanol (Double Volume) ice cold**—Lowers down the dielectric constant of the solution to precipitate out the DNA.
- **70% ethanol**—used for washing DNA as it lowers down the dielectric constant of the solution due to which salts get dissolved in the solution but DNA does remain insoluble.
- **7-M ammonium acetate**—furnishes NH^{3+} ions which bind to the phosphate backbone of DNA and reduces the repulsive negative force between two DNA strands and helps them to renature.
- **Chloroform:Isoamyl alcohol (24:1):** Mixing with chloroform:isoamyl alcohol followed by centrifugation separates proteins and other materials from DNA. Isoamyl alcohol forms a distinct zone in the middle that separates the aqueous phase from the chloroform phase. DNA will be present in the upper aqueous phase.
- **Tris-EDTA buffer (Tris-HCL = 10 mM (pH = 8.0), EDTA = 1 mM):** EDTA Chelates metal ions and helps in protecting DNA from nucleases. Maintains the pH of the solution for longer stability of DNA and storage.

Step-by-step method details

- **Extraction of DNA:**
- Homogenize in a mortar and pestle 300 mg of leaves by adding 500 μL of the CTAB buffer.

- Transfer the homogenized CTAB plant extract to a microfuge tube with the help of a clean spatula.
- **Lysis of DNA:**
- Incubate the sample at 65°C for 15 min in a rotating water bath.
- Spin the CTAB extract mixture at 12,000 g for 15 min after incubation, to spin down the cell debris.
- Transfer the supernatant to a new microfuge tube.
- **Phase separation and precipitation of DNA:**
- Add 250 µL of chloroform:Isoamyl alcohol (24:1) to each tube and mix the solution by inverting the tube 2–3 times.
- Spin the tubes at 13,000 g for 1 min.
- Transfer the upper aqueous phase, containing the DNA only, to a clean microfuge tube.
- To the tube, add ammonium acetate (7.5 M) @ 50 µL followed by chilled absolute ethanol @ 1 mL or chilled isopropanol @ 500 µL.
- Precipitate the DNA by inverting the tubes slowly several times. DNA precipitation can be seen in the form of a cottony white colored mass in the solution.
- Alternatively, add ethanol and place the tubes for 1 h at −20°C to enhance the precipitation of DNA.
- If the cottony mass does not appear, centrifuge the tube at 13,000 g for 15 min. Discard the supernatant.
- **Wash:**
- For washing the DNA, add ice-cold 70% ethanol @ 500 µL and slowly invert the tube and repeat (2 times).
- Centrifuge at 13,000 rpm for 1 min to form a pellet.
- Remove the supernatant.
- For the ethanol to evaporate, air-dry the pellet for 10–15 min under laminar flow.
- Resuspend the DNA in the Tris-EDTA buffer (volume depends on the size of the pellet) @ 100–200 µL.
- **Electrophoresis:**
- Analyze the integrity of the DNA via agarose gel electrophoresis in 0.7% agarose gel.

Expected outcomes

The different well lanes exhibit that purified DNA has been obtained with slightly visible RNA contamination when electrophoresed on agarose gel. RNA runs faster than gDNA on the gel owing to its lower molecular weight compared to DNA (Fig. 6).

Quantification and statistical analysis

DNA concentration can be determined spectrophotometrically at the given absorption using NanoDrop1000. The quality of the extracted DNA can be evaluated by electrophoresis separation for DNA samples on agarose gel stained with EtBr.

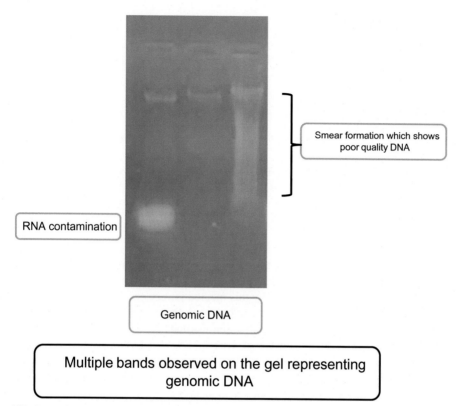

RNA contamination

Smear formation which shows poor quality DNA

Genomic DNA

Multiple bands observed on the gel representing genomic DNA

FIG. 6

Multiple bands observed on the gel representing genomic DNA.

Advantages

The technique is cheap, simple yet effective. It is the most reliable technique in the RAPD analysis of various plants and DNA barcoding.

Limitations

CTAB is hazardous to health. Phenol is volatile and can burn the skin.

Optimization and troubleshooting
Problem

Phenol is volatile and thus can evaporate rapidly. EtBr is carcinogenic and heat labile.

Potential solution

The bottle should be tightly closed after each use. Gloves should be worn while handling EtBr and the solution should be cooled down before mixing with EtBr.

Safety considerations and standards

- Always mix the samples gently and never vortex, in order to prevent sharing of DNA.
- Do not overdry the DNA or it will be difficult to redissolve.
- Do not add TE unless the smell of ethanol is gone.

Alternative methods/procedures

- **Genomic DNA extraction using the phenol-chloroform method (PCI)**

  PCI is one of the best methods of DNA extraction. If performed correctly, the yield and quality of DNA obtained by this method is very good. The method is also called the phenol-chloroform and isoamyl alcohol, PCI method of DNA extraction. The main chemicals for PCI DNA extraction methods are lysis buffer, phenol, and chloroform. The lysis buffer consists of Tris, EDTA, $MgCl_2$, NaCl, SDS, and other salts. In this method, the components of lysis buffer cause the lysis of the cell membrane as well as the nuclear envelope. The proteins of the cell are denatured using chloroform and phenol, which are organic in nature.

- **Salting out DNA extraction method**

  This method is safer compared to the PCI method. Salts such as sodium chloride, potassium acetate, and ammonium acetate are used to aid the DNA extraction. However, the inclusion of proteinase K causes the method to be aggressive. Use of different salts in DNA extraction may increase the yield but the purity would be low.

- **Enzymatic DNA extraction method**

  This method is in fact a combination of the salt method and enzymatic method. In this case, the extraction buffer is used before going further on enzymatic digestion. The major components of extraction buffer are: Tris, EDTA, NaCl, sodium lauryl, and SDS. Here the enzyme proteinase K is utilized for digesting the sample instead of phenol, chloroform, or isoamyl alcohol. The incubation of the sample with proteinase K for 2 h digests all the protein present inside the sample. Immediately after the proteinase K digestion, the sample is precipitated using chilled alcohol. All the remaining cell debris is removed by centrifuging the sample. Finally, the DNA pellet is dissolved in the TE buffer. This method of DNA extraction is rapid and easy. The yield is very high but the quality of DNA is a major concern.

- **Silica column-based DNA extraction method**

  The silica column-based DNA extraction method is very unique and differs from other DNA extraction methods. The method was first described by McCormick in *1989*. However, the idea was developed later in *1979*, when silica was used in DNA purification by Vogelstein. It works on the unique chemistry of interaction between silica and DNA. A positively charged silica particle binds with the negatively charged DNA and holds it during centrifugation. It is a widely accepted method due to the good yield and quality of DNA obtained. This method involves a simple operating system. The lysis buffer causes the cell membrane and the nuclear envelope to disintegrate and the proteinase K digests all the protein. The first step involves incubating the sample with a cell lysis buffer, also called a DNA extraction buffer. A small amount of proteinase k is also added to the sample. All the other impurities are removed by centrifugation. Here the DNA remains bound with silica and other impurities pass through the silica column. Now the DNA may be washed twice for the purpose of improving the purity in it. The impurities in the aqueous phase are discarded. Finally, the DNA is dissolved into the TE buffer. The method is fast, reliable, accurate, and less time-consuming as compared to other methods.

- **DNA extraction using anionic resins**

    In this method, the positively charged chelex binds to the negatively charged phosphate of DNA and helps in the extraction of DNA. The chelex is made up of the styrenedivinylbenzene copolymers. In this method, the column of the tube is filled with positive resins. The DNA binds to the positively charged resins and the cell lysate passes through the matrix. Proteins and other impurities or debris are washed off using the low concentration salt buffer. As a result, only the DNA remains in the matrix. The final step involves filling the matrix with the high concentration salt buffer which elutes the DNA from the resins. The DNA is then precipitated using the alcohol. The method is also called DNA extraction through anion exchange chromatography.

- **DNA extraction by magnetic beads**

    Positively charged magnetic beads attract the negatively charged DNA. The DNA is separated under the magnetic field. DNA extraction buffer is required for this technique as well.

- **CsCl density gradient method of DNA extraction**

    In this method, the DNA is separated based on its density using centrifugation. In high-speed centrifugation, the DNA band will appear at the isopycnic point, where the density of the DNA and the gradient (CsCl) become the same. This method is hard to perform as it requires high-speed centrifugation (10,000–12,000 rpm) for more than 10 h. Another major disadvantage of this method is the use of the carcinogenic EtBe in the DNA extraction. The EtBr intercalates between the strands of DNA and separates supercoiled DNA from the nonsupercoiled. This method is more suitable for the isolation of plasmid DNA.

References

Cheng, Y. J., Guo, W. W., Yi, H. L., Pang, X. M., & Deng, X. (2003). An efficient protocol for genomic DNA extraction from citrus species. *Plant Molecular Biology Reporter*, *21*(2), 177–178.

Porebski, S., Bailey, L. G., & Baum, B. R. (1997). Modification of a CTAB DNA extraction protocol for plants containing high polysaccharide and polyphenol components. *Plant Molecular Biology Reporter*, *15*(1), 8–15.

Sahu, S. K., Thangaraj, M., & Kathiresan, K. (2012). DNA extraction protocol for plants with high levels of secondary metabolites and polysaccharides without using liquid nitrogen and phenol. *International Scholarly Research Notices, 2012.*

Sambrook, J., & Russell, D. W. (2001). *Molecular cloning. A laboratory manual.* New York: Cold Spring Harbor Laboratory Press.

Genomic DNA extraction from silkworm using modified phenol:chloroform method (Suzuki et al., 1972)

Definition and rationale

In this method of extraction, the genomic DNA is isolated from the outer tissue of the fifth instar silkworm larvae using phenol:chloroform (Suzuki, Gage, & Brown, 1972). The extracted sample is first homogenized using a lysis buffer which contains EDTA that chelates magnesium ions, a necessary cofactor for nucleases. NaCl helps in the separation of polysaccharides from the DNA, as both DNA and polysaccharides have different solubility in CTAB based on the concentration of NaCl. At a higher concentration of NaCl, polysaccharides are insoluble, while at a lower concentration of NaCl, DNA is insoluble. Subsequently, by modifying the salt concentration, polysaccharides and DNA can be differentially precipitated. Then an equal volume of Tris saturated phenol is added and the sample is centrifuged. The supernatant is collected and an equal volume of phenol:chloroform:isoamyl (25:24:1) is added, mixed, and centrifuged to remove the protein contamination from the sample. The DNA extract is then treated with RNase to remove RNA contaminants. The DNA is then precipitated out and washed in organic solvents to remove contaminating salts before redissolving in aqueous solution and stored in chilled Tris-EDTA buffer and the quantification of DNA is done by running on a 0.7% agarose gel and/or taking absorbance at 260 nM. For further use, the extracted DNA should be stored at $-20°C$.

Before you begin

- DNA extraction from the outer tissue of the fifth instar silkworm larvae
- Preparation of gel matrix.

Key resources table

Reagent or resource	Source	Identifier
Tris	Sigma-Aldrich	MFCD00004679
EDTA	Sigma-Aldrich	MFCD00070672

Advanced Methods in Molecular Biology and Biotechnology. https://doi.org/10.1016/B978-0-12-824449-4.00008-6

Reagent or resource	Source	Identifier
NaCl	Sigma-Aldrich	MFCD00003477
SDS	Sigma-Aldrich	MFCD00036175
RNase	Thermo Fisher Scientific	AM2694
Phenol	Thermo Fisher Scientific	11,039,021
Chloroform	Sigma-Aldrich	MFCD00000826
Isopropanol	Sigma-Aldrich	MFCD00011674
Sodium acetate	Sigma-Aldrich	MFCD00012459
TrisEDTA buffer	Sigma-Aldrich	MFCD00236359

Materials and equipment

Materials and equipment required:

- Centrifuge
- Microfuge tubes (2 mL)
- Mortar and pestle
- Water bath (60–65°C)
- Sterile water
- Micropipettes
- Agarose electrophoresis equipment and reagents
- Shaking incubator

Reagents required:

- **200 mM Tris**—maintains pH.
- **25 mM EDTA**—It chelates metal ions such as Ca^{2+} and K^+ (important constituents of the cell membrane and keeps the membrane intact). Their chelation helps in disrupting the membrane. Also chelates Mg^{2+} ions, a necessary cofactor for nucleases, and prevents DNA degradation.
- **300 mM NaCl**—helps in the separation of polysaccharides from the DNA, as both of them have different solubility in CTAB depending on the concentration of NaCl. At a higher concentration of NaCl, polysaccharides are insoluble, while at a lower concentration of NaCl, DNA is insoluble. Subsequently, by modifying the salt concentration, polysaccharides and DNA can be differentially precipitated.
- **2% SDS**—a detergent that helps in lysis of the cell membrane.
- **RNase** @ 2 µg/mL
- **Phenol:chloroform:isoamyl (25,24:1)**—Helps in the removal of protein contamination.
- **Isopropanol (Equal Volume)/Absolute ethanol (Double Volume) ice cold**—Decreases the dielectric constant of the solution to precipitate out the DNA in the solution. Proteins and other materials are removed by adding

chloroform:isoamyl alcohol followed by centrifugation. Isoamyl alcohol forms a distinct zone in the middle that separates the aqueous phase from the chloroform phase.

- **70% ethanol**—used for washing DNA as it lowers down the dielectric constant of the solution due to which salts get dissolved in the solution but DNA remains insoluble.
- **3-M sodium acetate**—furnishes Na^+ ions which bind to the phosphate backbone of DNA and reduces the repulsive negative force between two DNA strands and helps them to renature.
- **Tris-EDTA buffer [10 mM Tris-HCL (pH = 8.0), 1 mM EDTA]**—Chelates metal ions and helps in protecting DNA from nucleases. Maintains the pH of the solution for longer stability of DNA and storage.

Step-by-step method details

- Genomic DNA shall be isolated from the outer tissue of the fifth instar silkworm larvae using the phenol:chloroform method (Suzuki et al., 1972).
- Homogenize 100 mg of the sample using 2 mL lysis buffer (200 mM tris, 25 mM EDTA, 300 mM NaCl, and 2% SDS) in a mortar and pestle.
- Add 1 μL proteinase K and 1 μL RNase and grind the tissue further. Then transfer it to a fresh microfuge tube and incubate at 60°C for 1 h in a water bath.
- Add equal volume of Tris saturated phenol and mix gently for 10 s.
- Then centrifuge the tubes at 2000 g for 10 min.
- Collect the supernatant in a new tube and add an equal volume of phenol:chloroform:isoamyl in the ratio of 25:24:1, mix and spin at 12,000 g for 10 min.
- Collect the upper aqueous layer in a fresh tube. For DNA precipitation, add 50 μL of 3 M sodium acetate and 2 volumes of absolute ethanol.
- Take out the spoolable DNA with the help of a pipette tip or inoculation loop and transfer into a fresh tube.
- Then wash the spooled DNA obtained with 70% ethanol and air-dry. Dissolve the DNA in a chilled Tris-EDTA buffer (10 mM Tris-HCl, 1 mM EDTA, pH − 8.0) and quantify it by running on 0.7% agarose gel and/or taking absorbance at 260 nM.
- For further use, store the extracted DNA at − 20°C.

Expected outcomes

The different lanes show that the purified DNA has been obtained with a slight RNA contamination when electrophoresed on agarose gel. Because of the lower molecular weight, RNA runs faster than g-DNA on the gel (Fig. 7).

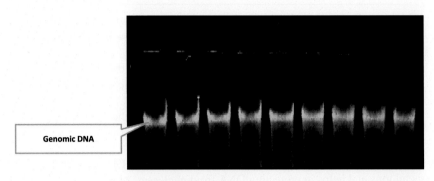

Genomic DNA

Gel showing DNA bands isolated from different strains of Silkworm

FIG. 7

Gel showing DNA bands isolated from different strains of silkworm.

Advantages

The technique is cheap, simple yet effective. It is the most reliable technique for DNA extraction.

Limitations

The phenol used in the technique is volatile and can burn the skin.

Optimization and troubleshooting
Problem

The phenol used in the technique is volatile. EtBr is a carcinogen and is heat labile nature.

Potential solution

The bottle in which the phenol is present should be tightly closed after each use. Gloves must be worn while handling EtBr and it should be mixed before the solution cools.

Safety considerations and standards

- Always mix the samples gently and never vortex, in order to prevent sharing of DNA.
- Do not overdry the DNA or it will be hard to redissolve
- Do not add TE unless the smell of ethanol is gone.

Alternative methods/procedures

- **Genomic DNA extraction using the phenol-chloroform method (PCI)**

  PCI is one of the best methods of DNA extraction. If performed correctly, the yield and quality of DNA obtained by this method is very good. The method is also called the phenol-chloroform and isoamyl alcohol, PCI method of DNA extraction. The main chemicals for PCI DNA extraction methods are lysis buffer, phenol, and chloroform. The lysis buffer consists of Tris, EDTA, $MgCl_2$, NaCl, SDS, and other salts. In this method, the components of lysis buffer cause the lysis of the cell membrane as well as the nuclear envelope. The proteins of the cell are denatured using chloroform and phenol, which are organic in nature.

- **Salting out DNA extraction method**

  This method is safer compared to the PCI method. Salts such as sodium chloride, potassium acetate, and ammonium acetate are used to aid the DNA extraction. However, the inclusion of proteinase K causes the method to be aggressive. Use of different salts in DNA extraction may increase the yield but the purity would be low.

- **Enzymatic DNA extraction method**

  This method is in fact a combination of a salt method and an enzymatic method. In this case, the extraction buffer is used before going further on enzymatic digestion. The major components of extraction buffer are: Tris, EDTA, NaCl, sodium lauryl, and SDS. Here the enzyme proteinase K is utilized for digesting the sample instead of phenol, chloroform, or isoamyl alcohol. The incubation of the sample with proteinase K for 2 h digests all the protein present inside the sample. Immediately after the proteinase K digestion, the sample is precipitated using chilled alcohol. All the remaining cell debris is removed by centrifuging the sample. Finally, the DNA pellet is dissolved in TE buffer. This method of DNA extraction is rapid and easy. The yield is very high but the quality of DNA is a major concern.

- **Silica column-based DNA extraction method**

  The silica column-based DNA extraction method is very unique and differs from other DNA extraction methods. The method was first described by McCormick in *1989*. However, the idea was developed later in *1979*, when silica was used in DNA purification by Vogelstein. It works on the unique chemistry of interaction between silica and DNA. A positively charged silica particle binds with the negatively charged DNA and holds it during centrifugation. It is a widely accepted method due to the good yield and quality of DNA obtained. This method involves a simple operating system. The lysis buffer causes the cell membrane and the nuclear envelope to disintegrate and the proteinase K digests all the protein. The first step involves incubating the sample with a cell lysis buffer, also called a DNA extraction buffer. A small amount of proteinase k is also added to the sample. All the other impurities are removed by centrifugation. Here the DNA remains bound with silica and other impurities pass through the silica column. Now the DNA may be washed twice for the purpose of improving the purity in it. The impurities in the aqueous phase are discarded. Finally, the DNA is dissolved into the TE buffer. The method is fast, reliable, accurate, and less time-consuming as compared to other methods.

- **DNA extraction using anionic resins**

  In this method, the positively charged chelex binds to the negatively charged phosphate of DNA and helps in the extraction of DNA. The chelex is made up of the styrene-divinylbenzene copolymers. In this method, the column of the tube is filled with positive resins. The DNA binds to the positively charged resins and the cell lysate passes through the matrix. Proteins and other impurities or debris are washed off using the low concentration salt buffer. As a result, only the DNA remains in the matrix. The final step involves filling the matrix with the high concentration salt buffer which elutes the DNA from the resins. The DNA is then precipitated using the alcohol. The method is also called DNA extraction through anion exchange chromatography.

- **DNA extraction by magnetic beads**

  Positively charged magnetic beads attract the negatively charged DNA. The DNA is separated under a magnetic field. DNA extraction buffer is required for this technique as well.

- **CsCl density gradient method of DNA extraction**

  In this method, the DNA is separated based on its density using centrifugation. In the high-speed centrifugation, the DNA band will appear at the isopycnic point, where the density of the DNA and the gradient (CsCl) become the same. This method is hard to perform as it requires high-speed centrifugation (10,000–12,000 rpm) for more than 10 h. Another major disadvantage of this method is the use of the carcinogenic EtBe in the DNA extraction. The EtBr intercalates between the strands of DNA and separates supercoiled DNA from the nonsupercoiled. This method is more suitable for the isolation of plasmid DNA.

References

Buhroo, Z., Ganai, N., & Malik, M. (2016). An efficient protocol for silk moth DNA extraction suitable for genomic fingerprinting and genetic diversity analysis. *Journal of Cell & Tissue Research, 16*(3).

Datta, R. K., & Ashwath, S. K. (2000). Strategies in genetics and molecular biology for strengthening silkworm breeding. *Indian Journal of Sericulture, 39*(1), 1–8.

Goldsmith, M. R., & Wilkins, A. S. (1995). Genetics of the silkworm: Revisiting an ancient model system. In *Molecular model systems in the lepidoptera* (pp. 21–76).

Singh, H. R., Unni, B. G., Neog, K., & Wann, S. B. (2011). Isolating silkworm genomic DNA without liquid nitrogen suitable for marker studies. *African Journal of Biotechnology, 10*(55), 11365–11372.

Suzuki, Y., Gage, L., & Brown, D. D. (1972). The genes for silk fibroin in *Bombyx mori. Journal of Molecular Biology, 70*, 637–649.

Genomic DNA extraction from Fungal Mycelium using the modified Cenis, 1992 method

Definition and rationale

- The modified Cenis method for genomic DNA isolation from fungal mycelium has been popularly used after its development. In this method, the fungal mycelium is first crushed in a lysis buffer as described by Cenis, 1992. EDTA is included in the extraction buffer to chelate magnesium ions, a necessary cofactor for nucleases to protect the released DNA from endogenous nucleases. Also, initial DNA extracts contain a large amount of polysaccharides, RNA, and proteins which are difficult to separate and may interfere with the DNA extraction. One method for purifying DNA using the extraction buffer exploits the differential solubility of polysaccharides and DNA in the buffer depending on the concentration of NaCl in the extraction buffer. At a higher concentration, polysaccharides are insoluble, while at a lower salt concentration, DNA is insoluble. This allows for the differential precipitation of polysaccharides and DNA by adjusting the salt concentration. Chloroform and isoamyl alcohol are used for the removal of proteins by denaturation and precipitation from the extract. RNA contamination, on the other hand, is normally removed by treating the extract with RNase.
- The DNA is precipitated and washed thoroughly in organic solvents so as to remove the contaminating salts. The purified DNA is then resuspended and stored in the Tris-EDTA buffer and is confirmed by gel electrophoresis.

Before you begin

1. DNA extraction from fungal mycelium.
2. Preparation of gel matrix.

Advanced Methods in Molecular Biology and Biotechnology. https://doi.org/10.1016/B978-0-12-824449-4.00009-8

Key resources table

Reagent or resource	Source	Identifier
Tris	Sigma-Aldrich	MFCD00004679
EDTA	Sigma-Aldrich	MFCD00070672
NaCl	Sigma-Aldrich	MFCD00003477
SDS	Sigma-Aldrich	MFCD00036175
RNase	Thermo Fisher Scientific	AM2694
Phenol	Thermo Fisher Scientific	11,039,021
Chloroform	Sigma-Aldrich	MFCD00000826
Isopropanol	Sigma-Aldrich	MFCD00011674
Sodium acetate	Sigma-Aldrich	MFCD00012459
TrisEDTA buffer	Sigma-Aldrich	MFCD00236359

Materials and equipment

Materials and equipment required:

- Centrifuge
- 2 mL microfuge tubes
- Mortar and pestle
- Water bath (60–65°C)
- Sterile water
- Micropipettes
- Agarose electrophoresis equipment and reagents.
- Shaking incubator

Reagents required:

- **200 mM tris pH 8.5**—maintains pH
- **25 mM EDTA**–EDTA chelates metal ions such as Ca^{2+} and K^+ (important constituents of the cell wall and keeps the wall intact). Their chelation helps in disrupting the cell wall and also prevents DNA degradation by chelating Mg^{2+} ions, a necessary cofactor for nucleases.
- **250 mM NaCl**—helps in the separation of polysaccharides from the DNA, as both of them have different solubility in CTAB depending on the concentration of NaCl. At a higher concentration of NaCl, polysaccharides are insoluble, while at a lower concentration of NaCl, DNA is insoluble. Subsequently, by modifying the salt concentration, polysaccharides and DNA can be differentially precipitated.
- **0.5% SDS**—is a detergent that helps in lysis of the cell membrane.
- **RNase @ 20 mg/mL**
- **Isopropanol (Equal Volume)/Absolute ethanol (Double Volume) ice cold**—decreases the dielectric constant of the solution to precipitate out

the DNA in the solution. Addition of chloroform:isoamyl alcohol followed by centrifugation removes proteins and other materials. Isoamyl alcohol forms a distinct zone in the middle that separates the aqueous phase from the chloroform phase.

- **70% ethanol**—used for washing DNA as it lowers down the dielectric constant of the solution due to which salts get dissolved in the solution but DNA remains insoluble.
- **Tris-EDTA buffer [10 mM Tris-HCl (pH = 8.0), 1 mM EDTA]**—chelates metal ions and helps in protecting DNA from nucleases. Maintains the pH of the solution for longer stability of DNA and storage.

Step-by-step method details

- DNA extraction can be done by using 400 μL of extraction buffer. This buffer is the same as that described by Cenis (1992), containing 200 mM Tris HCL (pH 8.5), 250 mM NaCl, 25 mM EDTA, and 0.5% SDS.
- The fungal mycelium is crushed in the extraction buffer using a mortar and pestle to make a slurry. The mixture is transferred to a sterilized 2-mL microfuge tube and incubated at 65°C in a water bath for 1 h.
- RNase @ 4 μL (20 mg/mL) is added to each tube and again incubated for 10 min at 65° with intermittent mixing.
- 3 M sodium acetate (130 μL), pH 5.2 is added, and the tube is incubated at −20°C for 10 min.
- The lysate is spun down at 4000 rpm at 4°C for 15 min and the supernatant is transferred to a clean microfuge tube.
- The DNA is precipitated out by adding 650 μL of isopropanol, pelleted at 4000 rpm for 10 min and then washed with chilled 70% ethanol to remove salt contaminants.
- Sample is air-dried for 30 min and resuspended in 100 μL of the Tris-EDTA buffer (pH 8.0).

Expected outcomes

The different well lanes show that purified DNA has been obtained with a slightly visible RNA contamination, when electrophoresed on agarose gel. Because of its lower molecular weight, RNA runs faster than gDNA on the gel (Fig. 8).

Advantages

The technique is simple and effective. It is a reliable technique for DNA barcoding.

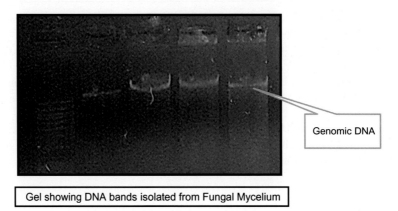

Genomic DNA

Gel showing DNA bands isolated from Fungal Mycelium

FIG. 8

Gel showing DNA bands isolated from Fungal Mycelium.

Limitations

The phenol used in the technique is volatile and can burn the skin.

Optimization and troubleshooting
Problem

Phenol is volatile in nature. EtBr is a carcinogen and heat labile.

Potential solution

The bottle should be tightly closed after each use. Gloves should be worn while handling EtBr and it should not be handled with bare hands. Phenol should not be inhaled or spilled on body.

Safety considerations and standards

- Always mix the samples gently and never vortex, in order to prevent sharing of DNA.
- Do not overdry the DNA or it will be hard to redissolve
- Do not add TE unless the smell of ethanol is gone.

Alternative methods/procedures

- **Genomic DNA extraction using the phenol-chloroform method (PCI)**

PCI is one of the best methods of DNA extraction. If performed correctly, the yield and quality of DNA obtained by this method is very good. The method is also called the phenol-chloroform and isoamyl alcohol, PCI method of DNA extraction. The main chemicals for PCI DNA extraction methods are lysis buffer, phenol, and chloroform. The lysis buffer consists of Tris, EDTA, $MgCl_2$, NaCl, SDS, and other salts. In this method, the components of lysis buffer cause the lysis of the cell membrane as well as the nuclear envelope. The proteins of the cell are denatured using chloroform and phenol, which are organic in nature.

- **Salting out DNA extraction method**

  This method is safer compared to the PCI method. Salts such as sodium chloride, potassium acetate, and ammonium acetate are used to aid DNA extraction. However, the inclusion of proteinase K causes the method to be aggressive. Use of different salts in DNA extraction may increase the yield but the purity would be low.

- **Enzymatic DNA extraction method**

  This method is in fact a combination of the salt method and enzymatic method. In this case, the extraction buffer is used before going further on enzymatic digestion. The major components of the extraction buffer are: Tris, EDTA, NaCl, sodium lauryl, and SDS. Here the enzyme proteinase K is utilized for digesting the sample instead of phenol, chloroform, or isoamyl alcohol. The incubation of the sample with proteinase K for 2h digests all the protein present inside the sample. Immediately after the proteinase K digestion, the sample is precipitated using chilled alcohol. All the remaining cell debris is removed by centrifuging the sample. Finally, the DNA pellet is dissolved in the TE buffer. This method of DNA extraction is rapid and easy. The yield is very high but the quality of DNA is a major concern.

- **Silica column-based DNA extraction method**

  The silica column-based DNA extraction method is very unique and differs from other DNA extraction methods. The method was first described by McCormick in *1989*. However, the idea was developed later in *1979*, when silica was used in DNA purification by Vogelstein. It works on the unique chemistry of interaction between silica and DNA. A positively charged silica particle binds with the negatively charged DNA and holds it during centrifugation. It is a widely accepted method due to the good yield and quality of the DNA obtained. This method involves a simple operating system. The lysis buffer causes the cell membrane and the nuclear envelope to disintegrate and the proteinase K digests all the protein. The first step involves incubating the sample with a cell lysis buffer, also called a DNA extraction buffer. A small amount of proteinase K is also added to the sample. All the other impurities are removed by centrifugation. Here the DNA remains bound with silica and other impurities pass through the silica column. Now the DNA may be washed twice for the purpose of improving the purity in it. The impurities in the aqueous phase are discarded. Finally, the DNA is dissolved in the TE buffer. The method is fast, reliable, accurate, and less time-consuming as compared to other methods.

- **DNA extraction using anionic resins**

  In this method, the positively charged chelex binds to the negatively charged phosphate of DNA and helps in the extraction of DNA. The chelex is made up of the styrene-divinylbenzene copolymers. In this method, the column of the tube is filled with positive resins. The DNA binds to the positively charged resins and the cell lysate passes through the matrix. Proteins and other impurities or debris are washed off using the low concentration salt buffer. As a result, only the DNA remains in the matrix.

The final step involves filling the matrix with the high concentration salt buffer which elutes the DNA from the resins. The DNA is then precipitated using the alcohol. The method is also called DNA extraction through anion exchange chromatography.

- **DNA extraction by magnetic beads**

Positively charged magnetic beads attract the negatively charged DNA. The DNA is separated under a magnetic field. DNA extraction buffer is required for this technique as well.

- **CsCl density gradient method of DNA extraction**

In this method, the DNA is separated based on its density using centrifugation. In high-speed centrifugation, the DNA band will appear at the isopycnic point, where the density of the DNA and the gradient (CsCl) become the same. This method is hard to perform as it requires high-speed centrifugation (10,000–12,000 rpm) for more than 10 h. Another major disadvantage of this method is the use of the carcinogenic EtBe in DNA extraction. The EtBr intercalates between the strands of DNA and separates the supercoiled DNA from the nonsupercoiled. This method is more suitable for the isolation of plasmid DNA.

References

Al-Samarrai, T. H., & Schmid, J. (2000). A simple method for extraction of fungal genomic DNA. *Letters in Applied Microbiology, 30*(1), 53–56.

Barbier, F. F., Chabikwa, T. G., Ahsan, M. U., Cook, S. E., Powell, R., Tanurdzic, M., et al. (2019). A phenol/chloroform-free method to extract nucleic acids from recalcitrant, woody tropical species for gene expression and sequencing. *Plant Methods, 15*(1), 62.

Cenis, J. L. (1992). Rapid extraction of fungal DNA for PCR amplification. *Nucleic Acids Research, 20*(9), 2380.

Das, S., & Dash, H. R. (2014). *Microbial biotechnology—A laboratory manual for bacterial systems.* Springer.

Raeder, U., & Broda, P. (1985). Rapid preparation of DNA from filamentous fungi. *Letters in Applied Microbiology, 1*(1), 17–20.

Bacterial transformation using CaCl$_2$

10

Definition

The process of receiving exogenous or foreign DNA by cells of interest that changes their phenotypic expressions is called *Transformation*. Before the uptake of foreign DNA, these host cells are made permeable to DNA and their efficacy to receive foreign DNA is increased. This state is termed *Competency*. Such chemically treated cells are termed competent cells. In nature, some bacteria have poor transforming ability or become incompetent due to environmental stresses, but they can purposely be made competent and their transforming potential could be increased by treating them with cold shock or giving treatment of chloride salts of metal cations such as calcium, etc. These treatments alter the structure and permeability of the cell wall and cell membrane of the bacterial cell so that the DNA can easily pass through the membrane and get incorporated into the cell. The number of cells transformed per 1 µg of DNA is called the *Transformation efficiency*. It ranges between 1×10^4 and 1×10^7 cells per µg of DNA added to the content.

Rationale

- Since DNA is a nonpolar and hydrophilic molecule of high magnitude, it cannot normally pass through the cell membranes of bacteria. Therefore, to ensure transformation, the host bacterial cells must be made competent to takeup the DNA. For this purpose, they are subjected to CaCl$_2$ treatment; the Ca^{2+} ions destabilize the cell membrane by perforating it and also form a complex with the exogenous DNA attaching cell surface. Incubating the competent cells and the foreign DNA together on ice followed by a brief heat shock aids the bacteria to take up the DNA. It has been found that *E. coli* Bacteria soaked in ice-cold solution followed by exposure to a temperature of 42°C are more efficient in the uptake of foreign DNA than untreated normal bacterial cells.
- The cations (Ca^{2+}) produced by the CaCl$_2$ salt form coordination complexes with the negatively charged DNA molecules. But DNA, being a macromolecule, cannot itself cross the membrane to enter into the cytosol of a host cell. The heat shock treatment as mentioned earlier strongly depolarizes the cell membrane of CaCl$_2$-treated cells. Due to this, there is a

Advanced Methods in Molecular Biology and Biotechnology. https://doi.org/10.1016/B978-0-12-824449-4.00010-4

decrease in the membrane potential that allows the movement of exogenous DNA into the host cell's interior. This is followed by the subsequent cold shock which raises the membrane potential to its previous original value. Another notion is that Ca^{2+} ions bind to the membrane and DNA, being negatively charged is attracted toward the membrane. Ca^{2+} ions also increase the pore size on the cell membrane so that extracellular DNA can easily pass through the membrane.

Before you begin

- Preparation of bacterial culture.
- Preparation of LB broth.

Key resources table

Reagent or resource	Source	Identifier
Antibodies		
NACl	Sigma-Aldrich	MFCD0003477
Triptone	Sigma-Aldrich	70,914
Yeast extract	Sigma-Aldrich	MFCD00132599
Agar	Sigma-Aldrich	MFCD00081288
Antibiotic	HIMEDIA	–
pEGFP-C1 plasmid		

Materials and equipment

Materials and equipment required:

- Centrifuge
- Eppendorf tubes
- Plasmid DNA
- Mortar and pestle
- Water bath

Reagents required:

- **LB broth:**
- 200 mM NaCl
- 1% Tryptone
- 0.5% Yeast extract
- **LB agar:**

Add 2% agar to LB broth prior to autoclaving.

- **M CaCl$_2$**
- **Antibiotic**—Any antibiotic gene that is on the plasmid that is to be transformed (in this case Kanamycin)
- **Plasmid**—pEGFP-C1

Step-by-step method details

Timing: Day 1: 4 h.

    Day 2: 30 min.

    All the below procedures should be strictly done under sterile conditions in a laminar airflow chamber. Everything should be autoclaved before use. The tubes, pipette tips, etc., that are to be used in the experiment should not be opened outside the laminar airflow.

1. Firstly, prepare a small overnight culture of the bacteria in LB broth medium. Grow this culture 37°C and avoid shaking the content.
2. Into 100 mL of fresh LB broth, inoculate 1.0 mL of the overnight culture. Grow this culture with rapid shaking at 37°C for 4 h.

    Note: Ideally, the bacteria should be grown at 37°C with continuous shaking; however, if not available, the bacteria will grow at room temperature and can be mixed manually every 10 min.

1. Spread 100 μL on two LB agar plates, one with an antibiotic and other without an antibiotic as negative and positive controls. Incubate the plates at 37°C overnight.
2. Take 1.5 mL of the aliquot and transfer it to a sterile centrifuge tube.
3. For 10 min, cool the content on ice.
4. After spinning for 5 min at 5000 g, pellet the cells.
5. Discard the supernatant carefully and resuspend cells in 0.5 mL of ice-cold 0.1 M CaCl$_2$.
6. For at least 20 min, leave it on ice and follow by tapping.
7. Then centrifuge it at 5000 g for 5 min.
8. It should be followed with resuspension of the cells in 200 μL of 0.1 M CaCl$_2$ (cold)
9. To each tube, add 1 μL (1 μg/μL concentration) of DNA.
10. Leave on ice (4°C) for 10 min.
11. Transfer the content to a 42°C water bath for 90 s and immediately return to ice for 5 more minutes.
12. Spread 100 μL on an LB agar plate devoid of antibiotic for control. This control is to check whether the cells have survived the heat shock. Incubate the plates at 37°C overnight.

13. It should be followed with the transfer of the contents of each tube to 500 µL of LB broth.
14. Incubate with shaking at 37°C for 2 h time.
15. Spin the cells at 5000 g for 5 min. Discard the major portion of the supernatant but keep 100 µL.
16. Resuspend the cells gently in this 100 µL residual supernatant.
17. Plate 100 µL aliquots onto an LB agar plate with an antibiotic. Incubate the plates at 37°C overnight.

Expected outcomes

Transformants that are harboring the plasmid DNA will form colonies on Kanamycin LB plates. The bacterial cells that are not able to transform plasmid into them will not grow on the Kanamycin LB plates because those will lack the resistance against Kanamycin. (Fig. 9).

Advantages

The technique is cheap and most reliable for bacterial transformation. It has a high transformation efficiency.

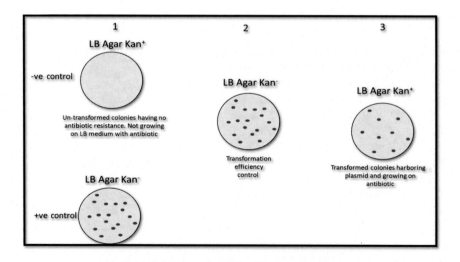

FIG. 9

Bacterial cell.

Limitations

Preparation of bacterial culture is a little cumbersome.

Optimization and troubleshooting
Problem

Bacterial cells are not always competent.

Potential solution

Harvest the cells which belong to a rapid growth phase.

Safety considerations and standards

- The bacterial cells when harvested for the said study should essentially belong to a rapid growth phase (exponential phase). They should not be allowed to reach the stationary phase.
- The competent cells can be stored at $-80°C$ only for a month for investigation.

Alternative methods/procedures

- Bacterial transformation using heat shock.
- Bacterial transformation using micro-shock waves.
- Bacterial transformation using a fluorescent protein.
- Bacterial transformation using PEG (polyethylene glycol).

References

Dagert, M., & Ehrlich, S. D. (1979). Prolonged incubation in calcium chloride improves the competence of *Escherichia coli* cells. *Gene, 6*(1), 23–28.

DeLucia, A. M., Six, D. A., Caughlan, R. E., Gee, P., Hunt, I., Lam, J. S., et al. (2011). Lipopolysaccharide (LPS) inner-core phosphates are required for complete LPS synthesis and transport to the outer membrane in *Pseudomonas aeruginosa* PAO1. *MBio, 2*(4).

Melcrová, A., Pokorna, S., Pullanchery, S., Kohagen, M., Jurkiewicz, P., Hof, M., et al. (2016). The complex nature of calcium cation interactions with phospholipid bilayers. *Scientific Reports, 6*, 38035.

Srivastava, S. (2013). *Genetics of bacteria*. Springer.

Stein, S. (1990). Production and analysis of proteins by recombinant DNA technology. In *Fundamentals of protein biotechnology* CRC Press.

Transformation of *E. coli* by electroporation

11

Definition

Another method for the transformation of bacteria, *Escherichia coli* is the electroporation method. In this method, the cells are allowed to grow up to the mid-log phase and are then extensively washed with water to remove all salts. Glycerol (10%) is added to the cells in order to freeze the cells for future use. *E. coli* cells are then mixed with the DNA to be transformed and then pipetted into a plastic cuvette containing electrodes. An electric pulse of 2400 V/am is given to the cells, which makes the membrane of the cells porous. The pores thus formed cause the entry of the DNA into the cells. These transformed cells are further kept on incubation with agar broth before they are plated. The effectiveness of transformation can be calculated as the number of transformants/µg of incorporated DNA. A negative control with cells that contain no added DNA should be included in the experiment for the purpose of comparison of degree of transformation.

Before you begin

- Preparation of bacterial culture.
- Preparation of LB broth.

Key resources table

Reagent or resource	Source	Identifier
Antibodies	HIMEDIA	–
NACl	Sigma-Aldrich	MFCD00003477
Ammonium acetate	Sigma-Aldrich	MFCD00013066
Triptone	Sigma-Aldrich	70,914
Yeast extract	Sigma-Aldrich	MFCD00132599
Agar	Sigma-Aldrich	MFCD00081288
Antibiotic	HIMEDIA	–
• *E. coli* host strain (DH5α)		
• Plasmid DNA		
• Trna		

Materials and equipment

Materials and equipment required:

- Centrifuge
- Centrifuge bottles (sterilized)—250 mL for GSA rotor
- Eppendorf tubes
- Plasmid DNA
- Mortar and pestle
- 500 mL of water bath for every 250 mL of culture (10% redistilled glycerol, 90% distilled water, *v*/v) kept at 4°C.
- Transformation plates
- Ethanol

Reagents required:

- LB broth:
- 200 mM NaCl
- Tryptone (1%)
- Yeast extract (0.5%)
- LB agar: Add agar (2%) to LB broth before autoclaving.
- 5 M ammonium acetate
- 0.5× TE or EB (10 mM Tris, pH 8.3)
- tRNA (5–10 µg/mL)—used as a mass carrier to increase the efficiency of precipitation
- SOB medium
- *E. coli* host strain, viz., DH5α

Step-by-step method details

1. **Preparation of *E. coli* cells for electroporation:**
   (a) A fresh colony of the suitable host strain (DH5α in this case) is used to inoculate 5 mL of SOB medium. The cells are grown overnight with vigorous aeration at 37°C.

   Note: SOB medium should be without magnesium

   (b) Besides 2.5 mL of cells are diluted in 250 mL of SOB in a flask having a capacity of 1 L. At 37°C, the cells are allowed to grow for 2–3 h with vigorous aeration until the cells reach an OD550 = 0.8.
   (c) The cells are harvested by centrifugation at 5000 rpm in sterile centrifuge bottles in a GSA rotor for 10 min
   (d) The cell pellets are then washed in 250 mL of chilled water bath:
   - A small amount of water bath is added to the cell pellet.
   - Pipette up and down or gently vortex until cells are resuspended.
   - Fill the centrifuge bottle with ice-cold water bath and gently mix.

(e) The cell suspension is again centrifuged at 5000 rpm for 15 min and then the supernatant is carefully poured off as soon as the rotor stops. Cells that have been washed in the water bath do not pellet well.

> **Note:** In case of turbidity of the supernatant, the centrifugation time needs to be increased.

(f) The cell pellet is washed for the second time by resuspending in 250 mL of the chilled water bath. The same technique is followed as defined above.

(g) The supernatant is carefully poured off, leaving a small amount of water bath at the bottom. The cell pellet is resuspended in the water bath and the final volume is kept at about 1 mL. The cells are either used immediately or frozen in 0.2 mL aliquots in freezer vials using a dry ice-ethanol bath. The frozen cells are stored at − 70°C.

2. Preparing DNA for electroporation:
   (a) DNA must have low ionic strength with a high resistance. It is purified by dilution, precipitation, or dialysis.
   (b) DNA is diluted in 10 mM Tris (pH 8–8.3) to about 1–50 ng/μL. About 1 μL of DNA is used for transformation.

3. Purifying DNA by precipitation:
   (a) In a 1.5 mL tube, 5–10 μg of tRNA is added to 20-μL ligation reaction, then 22 μL 5 M ammonium acetate (or an equal volume of ligation reaction with added tRNA) is added. The components are thoroughly mixed.
   (b) Also, 100-μL absolute ethanol (or 2.5 volumes of ligation reaction, tRNA and salt) is added. The mixture is chilled for 15 min.
   (c) The mixture is centrifuged at $> 12,000 \times g$ for 15 min at 4°C. The supernatant is carefully decanted.
   (d) The pellet is washed with 1 mL of ethanol (70%). It is then centrifuged at $> 12,000 \times g$ for 15 min at room temperature. The supernatant is then removed and the pellet is air-dried.
   (e) DNA in EB buffer (10 mM Tris-HCl, pH 8.3) or 0.5 × TE buffer [5 mM Tris-HCl, 0.5 mM EDTA (pH 7.5)] is resuspended to a concentration of 10 ng/μL of DNA. It is suitable to resuspend in 10 μL for ligation reactions; 1 μL is used for transformation of 20 μL of cell suspension.

4. Electroporation:
   (a) The required number of microfuge tubes is marked and the required number of micro-electroporation chambers are placed on the ice. The temperature control compartment of the chamber safe is filled with 250 mL of ice-cold water slurry. The chamber rack is placed in the chamber safe.
   (b) The aliquot of cells prepared for electroporation is thawed and 20 μL of the cells are added to the required number of microfuge tubes on ice. Then 1 μL of DNA or the ligation reaction is added to the microfuges.
   (c) Besides 20 μL of the cell DNA mixture is pipetted in the micro-electroporation chamber.

Note: Air bubbles should not be left in the droplet of cells because the pressure bubble may cause arcing and loss of the sample.

(d) The chamber is then placed in the slot in the chamber rack and its position is noted. The process is repeated if more than one sample is to be pulsed. Up to four samples can be placed in the chamber rack at one time. Chambers should be handled gently to avoid any accidental displacement of the sample from between the bosses.

(e) The chamber of the lid is closed safely and then secured by drawing the latch.

(f) The pulse cable is plugged into the right side of the chamber safe

(g) The chamber selection knob is turned on the top of the chamber safe to direct the electrical pulse to the desired micro-electroporation chamber.

(h) Resistance on the voltage booster is set to $4\,k\Omega$; pulse control unit to LOW and $330\,\mu F$. Connections should be double-checked.

(i) The pulse control unit is charged by setting the CHARGE ARM switch on the pulse control unit to CHARGE and then pressing the UP voltage control button until the voltage reading is 5–10 V higher than the desired discharge voltage. The standard conditions for *E. coli* are 2.4 kV, which means setting the pulse control unit to 405 V (400 V is the desired discharge voltage $+5$). The voltage booster amplifies the volts by ~ sixfold such that the total discharge voltage is 2400 V, or 2.4 kV. The actual peak voltage delivered to the sample will be shown on the voltage booster meter after the pulse is delivered.

(j) The CHARGE/ARM switch is set to ARM position. The green light indicates that the unit is ready to deliver a DC pulse. The pulse discharge trigger button is depressed and held for 1 s.

Note: The DC voltage display on the pulse control unit should read $< 10\,V$ after a pulse has been delivered. If not, discharge the capacitor using the DOWN button.

(k) The chamber selection knob is turned to the next desired and steps (h) and (i) are repeated until all the samples are pulsed.

(l) For ampicillin selection, the samples are inoculated into 2 mL of SOC medium and shaken for 30 min (for amp), 60 min (for Kan) to allow expression of the antibiotic gene. The cells are plated on LB medium with an appropriate antibiotic or screening reagent (e.g., 100 μg/mL ampicillin and/ or 40 μL of 20 mg/mL X-Gal, XP, and 40 μL of 100 mM IPTG).

Expected outcomes

Analyze what is seen on the plate; the section is compared with the control. Colonies should be present in all the sections, and no colonies can be seen in the control section. The difference is due to "Control" bacteria not having a plasmid, as the plasmid confers antibiotic resistance to the bacteria that were growing on the antibiotic plate.

Advantages

This technique is cheap and has quite good transformation efficiency.

Limitations

Preparation of bacterial culture is a little cumbersome.

Optimization and troubleshooting
Problem

Bacterial cells are not always competent.

Potential solution

Harvest the cells which belong to a rapid growth phase.

Safety considerations and standards

- The bacterial cells when harvested for the said study should essentially belong to a rapid growth phase (exponential phase). They should not be allowed to reach the stationary phase.
- The competent cells can be stored at $-80°C$ only for a month for investigation.

Alternative methods/procedures

- Bacterial transformation using $CaCl_2$
- Bacterial transformation using heat shock.
- Bacterial transformation using micro-shock waves.
- Bacterial transformation using a fluorescent protein.
- Bacterial transformation using PEG (polyethylene glycol).

Further reading

Sambrook, J., & Russell, D. W. (2006). Transformation of *E. coli* by electroporation. *Cold Spring Harbor Protocols*, *2006*(1). pdb-prot3933.

Weaver, J. C. (1995). Electroporation theory: Concepts and mechanisms. In J. A. Nickollof (Ed.), *Animal cells electroporation and electrofusions protocols*. Totowa, NJ: Humana Press.

Wu, N., Matand, K., Kebede, B., Acquaah, G., & Williams, S. (2010). Enhancing DNA electrotransformation efficiency in *Escherichia coli* DH10B electrocompetent cells. *Electronic Journal of Biotechnology*, *13*(5), 21–22.

Transfection of cells using lipofectamine

12

Definition

Transfection is a technique that introduces foreign nucleic acids into the host cells, thus producing genetically modified cells. The method helps us to study the function and regulation of genes and proteins. The introduced genetic materials exist in cells either in a stable condition or transiently depending on the nature of the genetic materials (Kim & Eberwine, 2010). For transfection to be stable, the introduced genetic material is integrated into the host genome and sustains transgene expression even after host cells replicate (Guo, Guo, Xu, Jia, & Jia, 2019). On the other hand, transiently transfected genes are only expressed for a limited period of time and are not integrated into the genome (Gelbart, Knobler, Garmann, Azizgolshani, & Cadena-Nava, 2019). These can be lost due to the cell division or environmental factors; therefore, the choice of stable or transient transfection depends on the aim of the experiment. The gene and gene products can be studied by increasing or inhibiting specific gene expression in cells (Kumar, Nagarajan, & Uchil, 2019). The method of transfection is classified into biologically, chemically, and physically mediated ones.

The methods most widely used in modern research are the chemical methods. These methods were the first to be used for the introduction of foreign genes into mammalian cells. Chemical transfection methods are most widely used in contemporary research and are known to be the first used method for the introduction of foreign genes into mammalian cells (Liu et al., 2020). These methods make use of chemical reagents such as Lipofectamine-2000, Lipofectamine-3000, and Profectamine. The principle behind the method is that positively charged chemical reagents make nucleic acid/chemical complexes with negatively charged nucleic acids. These positively charged nucleic acid/chemical complexes are attracted to the negatively charged cell membrane (Sokolova, Rojas-Sánchez, Białas, Schulze, & Epple, 2019). Endocytosis and phagocytosis are the processes that are believed to play a role in the passing of these complexes through the cell membranes. For the genes to get expressed, the transfected DNA must be delivered to the nucleus. The translocation of the complexes to the nucleus remains unknown. Chemical methods of transfection have merits of relatively low cytotoxicity, no mutagenesis, no extra carriage of DNA, and no size limitation on the packaged nucleic acid (Olbrich et al., 2019).

Transfection of mammalian cells in labs requires the maintenance of cells routinely in growth medium (e.g., RPMI-1640 medium), supplemented with 10% FBS, 1% L-glutamine, and 100 μg/mL penicillin/streptomycin in a 5% CO_2 incubator at

Advanced Methods in Molecular Biology and Biotechnology. https://doi.org/10.1016/B978-0-12-824449-4.00012-8

37°C. The reagents used in growth media provide an appropriate culturing condition such as pH, nutrients, substrate, growth factors, and hormones and prevent contamination of cell cultures.

A protocol that is generally followed for transfection of mammalian cells in labs involves growing the cells to 70%–90% confluency by seeding in 6/12/24 well plates. On the day of transfection, 1 h prior to transfection, growth media such as RPMI should be removed from cells and OptiMEM should be added to maintain serum-free conditions, as the albumin component of the serum limits the ability of the transfection reagent to cause transfection. At the time of transfection, both the transfection reagent such as Lipofectamine and nucleic acids should be diluted in OptiMEM media. Then diluted nucleic acids should be added to diluted transfection reagents such as Lipofectamine and incubated for 20 min. Nucleic acid/lipid complexes should then be added to the cells and maintained in a CO_2 incubator. After 16 h of transfection, cells should be visualized under a fluorescent microscope to check for transfection efficiency.

Key resources table

Reagent or resource	Source	Identifier
FBS	Sigma-Aldrich	S0615
PBSI	Sigma-Aldrich	
Lipofectamine	Sigma-Aldrich	L3287

Materials and equipments

Reagents required:

- 10% FBS in OptiMEM
- 1 × PBS
- OptiMEM
- Lipofectamine-2000

Step-by-step method details

The procedure is for 10 cm plates

- Before using FBS for the experiment, it should be thawed. Make 10% FBS in filter sterilized OptiMEM media containing no antibiotics.
- Aspirate the medium from the cells
- Wash the cells with 2–3 PBS
- Aspirate off PBS

- Add 10% FBS/OptiMEM media/plate
- Incubate the plates at 37°C for 30 min while mixing
  A Tubes: Selected DNA and QS with OptiMEM (100 μL)
  B Tubes: Lipofectamine-2000 and QS with OptiMEM (100 μL)
  Incubate for 5 min
- Add the components of tube B to tube A by bubble mixing. Incubate the complex for 30 min at room temperature.
- Pour 2001-μL A/B complex mixture into the cells, respectively. Gently mix the media by moving back and forth. Incubate in hood at 37°C for 6 h.
- Remove OptiMEM media containing A/B complexes.
- Add 8 mL plate of 10% FBS/DMEM with antibiotics.

Advantages

- The nucleic acid/Lipofectamine-2000 complexes can be added directly to cells in the presence and absence of the serum.
- The complexes need not be removed/changed/added after transfection but can be removed after 4–6 h

Optimization and troubleshooting
Problem

There is a chance of cytotoxicity.

Potential solution

Cultures should be transfected at a higher confluence and use less Lipofectamine and DNA to avoid cytotoxicity.

References

Gelbart, W. M., Knobler, C. M., Garmann, R. F., Azizgolshani, O., & Cadena-Nava, R. D. (2019). *In vitro reconstituted plant virus capsids for delivering RNA genes to mammalian cells: Google Patents.*

Guo, J., Guo, C., Xu, L., Jia, J., & Jia, R. (2019). Enhanced transfection efficiency by using a novel semi-attachment method in cell line and primary cells. *Analytical Biochemistry, 587,* 113465.

Kim, T. K., & Eberwine, J. H. (2010). Mammalian cell transfection: The present and the future. *Analytical and Bioanalytical Chemistry, 397*(8), 3173–3178.

Kumar, P., Nagarajan, A., & Uchil, P. D. (2019). DNA transfection mediated by cationic lipid reagents. *Cold Spring Harbor Protocols, 2019*(3) (pdb. prot095414).

Liu, W., Rudis, M. R., Cheplick, M. H., Millwood, R. J., Yang, J.-P., Ondzighi-Assoume, C. A., et al. (2020). Lipofection-mediated genome editing using DNA-free delivery of the Cas9/gRNA ribonucleoprotein into plant cells. *Plant Cell Reports*, *39*(2), 245–257.

Olbrich, T., Vega-Sendino, M., Murga, M., de Carcer, G., Malumbres, M., Ortega, S., et al. (2019). A chemical screen identifies compounds capable of selecting for Haploidy in mammalian cells. *Cell Reports*, *28*(3), 597–604. e594.

Sokolova, V., Rojas-Sánchez, L., Białas, N., Schulze, N., & Epple, M. (2019). Calcium phosphate nanoparticle-mediated transfection in 2D and 3D mono-and co-culture cell models. *Acta Biomaterialia*, *84*, 391–401.

Total RNA isolation

13

Definition and rationale

Total RNA isolation is a method based on liquid-phase separation which helps to obtain pure RNA in aqueous phase. The RNA obtained is precipitated, dissolved, and reprecipitated followed by washing in alcohol. The RNA sequestrated has many other sensitive applications such as sequencing studies, RNase protection assays, or reverse transcription-PCR. Enzymes present in the live cells are able to degrade RNA even after removal of the tissue sample from the organism. In order to prevent this degeneration, the samples must be immediately frozen at $-80°C$ until the RNA can be extracted. For the extraction purpose, the frozen tissues/cells should be immersed in the denaturing solution and rapidly homogenized before the tissue thaws for the purpose of inactivation of RNase and thus prevention of RNA degradation. The balance of salt concentration and pH determines the quality of the recovered RNA. The extraction solution overloaded with tissue or usage of the denaturing beyond the prescribed amount can also be responsible for the degradation of the quality of the RNA. Overdrying of the RNA pellets can also obstruct RNA solubilization. RNA should be solubilized at a concentration that will be appropriate for meaningful spectrophotometric quantitation as well as subsequent downstream molecular biology applications. During RNA solubilization and storage, care needs to be taken to avoid RNase contamination.

The level of the endogenous RNase varies with the cell type, and thus the necessary precautions may also vary. These may involve the use of guanidinium thiocyanate (GuSCN), phenol, a thiol reagent (β-mercaptoethanol and dithiothreitol), proteinase K, a detergent (sodium dodecyl [lauryl] sulfate, N-dodecyl sarkosine [sarkosyl]), placental RNase inhibitor, and vanadyl ribonucleoside complexes. After the addition of the GuSCN, RNA may be separated from protein and DNA using phenol prior to precipitation with ethanol or isoproponol. The pH should be maintained at 7.0 ± 0.2. Phenol treatment may also be given. Chloroform alone, without phenol, has been recommended for bacterial RNA extraction. Polyethylene glycol is used to precipitate intact virus particles and plasmid DNA during plasmid or viral DNA preparation.

Advanced Methods in Molecular Biology and Biotechnology. https://doi.org/10.1016/B978-0-12-824449-4.00013-X

Before you begin

Autoclave the solutions before beginning with 0.1% (*v*/v) DEPC.

Key resources table

Reagent or resource	Source	Identifier
SDS	Sigma-Aldrich	MFCD00036175
GuSCN		
Sodium citrate	Sigma-Aldrich	MFCD00150031
• β-Mercaptoethanol	Sigma-Aldrich	MFCD00004890
Proteinase K	Sigma-Aldrich	MFCD00132129
NaCl	Sigma-Aldrich	MFCD00003477
PEG	Sigma-Aldrich	MFCD00081839
Sodium phosphate buffer	Sigma-Aldrich	MFCD00149182
Tris-EDTA buffer	Sigma-Aldrich	MFCD00236359
Agarose	Sigma-Aldrich	MFCD00081294
LB broth	Sigma-Aldrich	L3397

Materials and equipment

Materials and equipment required:

- Gel casting apparatus (glass plates and spacers)
- Comb
- Clamps
- Micropipettes
- Conical flasks
- Spectrophotometer
- Spatula
- Distilled water
- Sample

Reagents required:

- SDS 5% (*w*/v).
- Guanidinium thiocyanate (GuSCN) 6 M. Filter and store at room temperature.
- 4-M guanidinium thiocyanate.
- 25-mM sodium citrate, pH 7.0.
- β-Mercaptoethanol.
- 0.5 mg/mL. Proteinase K: Aliquot (store at −20°C).
- 0.85% (*w*/v) NaCl physiological saline. Autoclave and store at room temperature.

- 20% (*w/v*) PEG 6000, 0.75% (*w/v*) SDS in 7.5% (w/v) potassium phosphate, pH 7.2. Autoclave and store at room temperature.
- 0.12-M sodium phosphate buffer, pH 7.2.
- Phenol saturated with 0.12-M phosphate buffer, pH 7.2. Add phosphate buffer to the previously opened bottle of phenol (either freshly redistilled at 165–180°C, or provided for use in molecular biology) in a previously unopened bottle. Mix and add 0.05% (*w*/w of phenol) 8-hydroxyquinoline as an antioxidant and allow phase separation at 4°C. Remove and discard the upper aqueous layer and repeat the buffer addition, mixing and decanting phenol. Store phenol at − 20°C.
- Chloroform.
- Sodium acetate, 3 M, pH 5.2.
- TBE buffer: Tris borate EDTA buffer: 90-mM Tris (hydroxymethyl) aminomethane, 90-mM boric acid, 2.5 mM EDTA, pH 8.3. Prepare a stock 10× solution by dissolving 108-g Tris base, 55-g boric acid, and 9.5-g disodium EDTA in water, make the final volume to 1 L.
- Agarose, 1.1% (*w/v*) in TBE. Agarose, 1.0% (*w/v*). Agarose should be low electroendosmosis (EEO).
- TE buffer: 10 mM Tris-HCl, pH 8.0, 1 mM EDTA. Autoclave.
- 17.10% (*w/v*) sodium dodecyl sulfate. Autoclave.
- L-broth: 10 g/L tryptone, 5 g/L yeast extract, 5 g/L NaCl, 1 g/L glucose

---

Step-by-step method details

Timing: 4 h

1. **Bacterial culture**
    (a) Inoculate 50 mL of LB broth and incubate overnight at 37°C until the absorbance reaches to 0.45–0.60 at 600 nm.
    (b) Centrifuge a 1.5 mL sample at 12,000 rpm for 10 min. The supernatant is discarded.
    (c) Wash in saline by resuspending the pellet in 300 μL of saline; centrifuge at 12,000 rpm for 10 min and discard the supernatant.
    (d) Resuspend the pellet in 400 μL of 0.12-M sodium phosphate buffer having a pH of 7.2
2. **Lysis and extraction with PEG 6000**
    (a) Add 50 μL of 5% SDS and 50 μL of proteinase K (0.5 mg/mL) to the resuspended pellet. Vortex and incubate at 37°C for 20 min.
    (b) Add 500 μL of PEG 6000 (12.5% *w/v*) to 1% potassium phosphate, pH 7.2. Vortex and centrifuge at 12,000 rpm for 10 min.
    (c) Remove the supernatant carefully, measure the volume into the sterile tube for precipitation of RNA. Phenol extraction can be done at this stage if required.

(d) Add 0.1 volume of 3-M sodium acetate, pH 5.2, followed by 2.5 volume of ethanol (or 1 volume of isopropanol) at $-20°C$ to precipitate the RNA. Keep it as such for an hour

(e) Centrifuge it at 12,000 rpm for 15 min, throw away the supernatant, and drain the tube by inverting on a tissue paper. Wash the pellet with 70% ethanol resuspending and centrifuging at 12,000 rpm for 10 min. Decant and drain.

(f) Dissolve the pellet in TE at room temperature for at least 30 min. The yield is about 15-μg total RNA/mL of culture. The ratio A260/A280 is in the range 1.90–2.05.

(g) Add 0.1 volume of buffer-saturated phenol for protein removal. Vortex and centrifuge at 12,000 rpm for 10 min. Separate the upper aqueous layer, repeat the extraction of this layer with 1 volume of phenol, vortex, and centrifuge, and retain the upper layer. The RNA may be precipitated without adding more sodium acetate.

3. **Lysis and extraction with GuSCN**

   (a) Addition after Cell Lysis

   - Cells are prepared and lysed.
   - Add 75 μL of 6 M GuSCN, vortex and centrifuge at 12,000 rpm for 10 min. Transfer the supernatant to a fresh tube.
   - RNA can be precipitated directly at this stage, without significant protein contamination, by the addition of 0.1 volume of sodium acetate and 2.5 volumes of cold ethanol.
   - Alternatively, phenol extraction may be carried out prior to precipitation.

   (b) Addition before cell lysis

   - In the presence of GuSCN, bacteria can be lysed by brief sonication. This procedure is just a one-step acid GuSCN-phenol method, except that the detergent is added after sonication in order to avoid froth.
   - The bacterial pellet is resuspended in 1 mL of GuSCN (4 M), sodium citrate, β-mercaptoethanol and sonicated for 20 s.
   - Then 50 μL of 10% sarkosyl is added, given the vortex, and centrifuged at 12,000 rpm for 10 min.
   - The RNA may be precipitated at this particular stage with the addition of 0.1 volume of 2 M sodium acetate, pH 4.0 and either 2.5 volume of ethanol or 1 volume of isopropanol at $-20°C$ followed by centrifugation at 12,000 rpm for 10 min, or phenol-extracted in the presence of acidic sodium acetate and acidic phenol.
   - In addition 0.1 volume of 2-M sodium acetate, pH 4.0, 1 volume of water, saturated phenol, and 0.2 volume of chloroform isoamyl alcohol (49:1, *v/v*) are subsequently added, then made to vortex and centrifuged at 12,000 rpm for 10 min.
   - The RNA in the upper aqueous phase is precipitated by the addition of 2.5 volumes of ethanol or 1 volume of isopropanol.

The procedure is diagrammatically explained in Fig. 10.

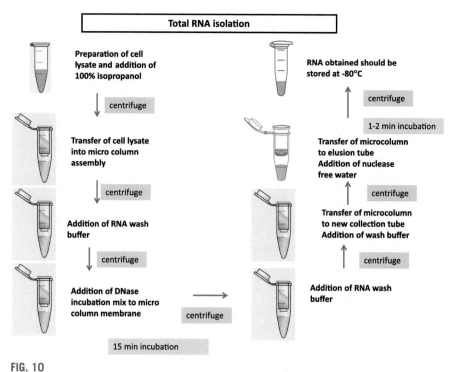

Total RNA isolation

Preparation of cell lysate and addition of 100% isopropanol

centrifuge

Transfer of cell lysate into micro column assembly

centrifuge

Addition of RNA wash buffer

centrifuge

Addition of DNase incubation mix to micro column membrane

15 min incubation

centrifuge

Addition of RNA wash buffer

centrifuge

Transfer of microcolumn to new collection tube Addition of wash buffer

centrifuge

Transfer of microcolumn to elusion tube Addition of nuclease free water

1-2 min incubation

centrifuge

RNA obtained should be stored at -80°C

FIG. 10

Total RNA isolation.

Expected outcomes

Precipitated RNA is obtained in the tube which should be stored at $-80°C$.

Advantages

The technique is versatile and can be compatible with a variety of sample types. This technique can process small and large samples. It is an inexpensive procedure and is a well-established and proven technology.

Limitations

Not high-throughput. RNA may contain contaminating genomic DNA.

Optimization and troubleshooting
Problem

RNA degradation.

Potential solution

Tissues should be immediately processed or 'snap-frozen' in liquid nitrogen to avoid RNase contamination. The solutions should be treated with DEPC; sterile RNase-free glassware and plastic ware should be used; clean gloves should be worn and RNA should be dissolved in DEPC-treated water stored at $-5°C$ to $-20°C$. RNA should be stored at $-80°C$.

Safety considerations and standards

- Handle all the reagents involved with gentle care.
- Care should be taken while handling GuSCN and SDS, particularly when in the solid state.
- DEPC is a carcinogen, and thus should be handled with care.
- Phenol burns the skin. If this occurs, wash with 20% (*w/v*) PEG in 50% (*v/v*) industrial methylated spirits.

Alternative methods/procedures

- Guanidinium-acid-phenol extraction.
- Silica technology, glass fiber filters.
- Density gradient centrifugation using cesium chloride or cesium trifluoroacetate.
- Magnetic bead technology.
- Lithium chloride and urea isolation.

References

Chirgwin, J. M., Przybyla, A. E., MacDonald, R. J., & Rutter, W. J. (1979). Isolation of biologically active ribonucleic acid from sources enriched in ribonuclease. *Biochemistry*, *18*(24), 5294–5299.

Chomczynski, P., & Sacchi, N. (2006). The single-step method of RNA isolation by acid guanidinium thiocyanate–phenol–chloroform extraction: Twenty-something years on. *Nature Protocols*, *1*(2), 581–585.

Clark, M. S. (Ed.). (2013). *Plant molecular biology—A laboratory manual* Springer Science & Business Media.

Puissant, C., & Houdebine, L. M. (1990). An improvement of the single-step method of RNA isolation by acid guanidinium thiocyanate-phenol-chloroform extraction. *BioTechniques*, *8*, 148–149.

Sambrook, J., Fritsch, E. F., & Maniatis, T. (1989). *Molecular cloning: A laboratory manual* (2nd ed.). Cold Spring Harbor Laboratory Press.

RNA isolation from plant tissues with high phenolic compounds and polysaccharides

14

Definition and rationale

RNA is rich in secondary metabolites such as phenolic compounds and polysaccharides that coprecipitate with nucleic acids. Therefore, isolation of good quality RNA from plant tissues is a challenging task. These phenolic compounds lead to the oxidation and degradation of RNA, thus making it unfit to conduct further experiments. In this technique, RNA is isolated from the leaves of the plants. Here polyvinylpyrrolidone (PVP) is added in the extraction buffer, so that it can bind to the phenolic compounds which are later eliminated by ethanol precipitation.

Before you begin

- Preparation of Gel matrix.
- Preparation of samples for running in the gel.
- Preparation of RNA samples.

Key resources table

Reagent or resource	Source	Identifier
Liquid nitrogen	Sigma-Aldrich	Z150851
EDTA	Sigma-Aldrich	–
NaCl	Sigma-Aldrich	–
Tris-HCl	Sigma-Aldrich	–
Sodium dodecyl sulfate (SDS)	Sigma-Aldrich	–
PVP	Sigma-Aldrich	–
Chloroform:isoamyl alcohol (CI, 24:1 *v/v*)	Sigma-Aldrich	C0549
Phenol:chloroform isoamyl alcohol (PCI, 1:1 v/v)	Sigma-Aldrich	MFCD00133763
Sodium acetate	Sigma-Aldrich	MFCD00012459

Advanced Methods in Molecular Biology and Biotechnology. https://doi.org/10.1016/B978-0-12-824449-4.00014-1

Materials and equipment

Materials and equipment required:

- Gel casting apparatus
- Comb
- Clamps
- Micropipettes
- Conical flasks
- Spectrophotometer
- Spatula
- Distilled water
- RNA sample
- Precooled pestle and mortar
- High-speed centrifuge
- Spectrophotometer
- Vortex mixer

Reagents required:

- Liquid nitrogen
- Extraction buffer: NaCl (0.25 M), Tris-HCl (0.05 M), (pH 7.5); EDTA (20 mM), 1% (*w/v*) sodium dodecyl sulfate (SDS), and 4% (*w/v*) PVP
- Chloroform:isoamyl alcohol (24:1 *v/v*)
- Phenol:chloroform isoamyl alcohol (1:1 *v/v*)
- 3-M sodium acetate (adjusted to pH 5.2 with acetic acid)
- Cold 70% ethanol (*v/v*)
- Cold absolute ethanol
- 0.1% diethyl pyrocarbonate (DEPC)—(*v/v*) treated-autoclaved water
- LiCl (10 M)
- Ladder

Step-by-step method details

Timing: steps 3–5 h.

1. **RNA extraction**
   (a) A total of 7.5 mL of extraction buffer and 7.5 mL of CI are added to a 30 mL eppendorf tube.

Critical step
   (b) A 1-g frozen leaf sample is ground to fine powder with a mortar and pestle in liquid nitrogen.
   (c) A 1 g of ground sample is transferred to a tube containing the extraction buffer and CI. Vortex the mixture vigorously and centrifuge at $12,857 \times g$ for 2 min at room temperature.

(d) The supernatant is transferred to a new 30-mL eppendorf tube and purified with an equal volume of PCI.

(e) It is then centrifuged at $12,857 \times g$ for 2 min at room temperature.

(f) This step is repeated until a clean interface is observed.

Critical step

(g) The supernatant is transferred to another new 30 mL eppendorf tube and one-tenth of the volume of 3-M sodium acetate having pH 5.2 and 2.5 volume of cold absolute ethanol are added, mixed well, and incubated at 4°C for 30 min.

Critical step

(h) The nucleic acids are recovered by centrifugation at $12,857 \times g$ for 20 min at 4°C.

(i) The pellet thus formed is washed with 70% (v/v) ethanol, air-dried, and 200 μL DEPC-treated water is later added to dissolve the pellet.

(j) The supernatant is then transferred to a 1.5 mL tube and 10 M LiCl is added to the final concentration of 2 M. It is then kept on ice for 1 h. Centrifuge at $18,514 \times g$ for 20 min at 4°C.

(k) The pellet is then washed with 70% (v/v) ethanol, air-dried, and 20-μL DEPC-treated water is added to dissolve the pellet. Centrifuge at $18,514 \times g$ for 10 min at 4°C.

(l) A 0.1 volume of 3-M sodium acetate having pH 5.2 and 2.5 volume of cold absolute ethanol are added to it, mixed well, and then stored at $-80°C$.

Expected outcomes

The different lanes demonstrate that purified RNA has been obtained when electrophoresed on agarose gel. RNA has lower molecular weight, and thus runs faster on the gel.

Quantification and statistical analysis

The total RNA is quantified with a spectrophotometer. The integrity of total RNA can be estimated by analyzing approximately 1-μg RNA sample on 1% (w/v) formaldehyde denaturing agarose gel.

Advantages

The technique is cheap, simple yet effective. This is the rapid method of RNA isolation from plant tissues.

Limitations

Phenol is volatile and can burn the skin.

Optimization and troubleshooting
Problem
Phenol is volatile, and thus can evaporate rapidly. EtBr is carcinogenic and heat labile.

Potential solution
The bottle should be tightly closed after each use. Gloves should be worn while handling EtBr and the solution should be cooled down before mixing with EtBr.

Safety considerations and standards
- Always mix the samples gently and never vortex, in order to prevent sharing of RNA.
- Do not overdry the RNA or it will be hard to redissolve.

Further reading

Dabo, S. M., Mitchell, E. D., & Melcher, U. (1993). A method for the isolation of nuclear DNA from cotton (Gossypium) leaves. *Analytical Biochemistry*, *210*(1), 34–38.

Loomis, W. D. (1974). Overcoming problems of phenolics and quinones in the isolation of plant enzymes and organelles. In *Vol. 31. Methods in enzymology* (pp. 528–544). Academic Press.

Matsumura, H., Nirasawa, S., & Terauchi, R. (1999). Transcript profiling in rice (*Oryza sativa* L.) seedlings using serial analysis of gene expression (SAGE). *The Plant Journal*, *20*(6), 719–726.

Rochester, D. E., Winer, J. A., & Shah, D. M. (1986). The structure and expression of maize genes encoding the major heat shock protein, hsp70. *The EMBO Journal*, *5*(3), 451–458.

Sambrook, J., Fritsch, E. F., & Maniatis, T. (1989). *Molecular cloning: A laboratory manual* (2nd ed.). Cold Spring Harbor Laboratory Press.

Total protein extraction with TCA-acetone and 2D gel electrophoresis

15

Definition

Proteome may be defined as the sum of all proteins produced by the cell under any specified condition with respect to tissue or cell types. Proteins are the main workforce of the cell and in eukaryotic systems, although DNA remains constant in all the tissues, the actual number of proteins produced by genes remains a mystery, as none of the methods of proteomics could measure the total number of proteins. With respect to the plant, each tissue has its typical characteristics depending on the role it plays; for instance, leaf tissues show very little problem during proteomics studies because of a lesser number of interferents or nonprotein components than root tissues. These reduce the numbers of proteins visualized under two-dimensional gel electrophoresis. Different protocols are assigned for different plant tissues to optimize protein extraction for each tissue.

A comparison of 2D gels requires well-resolved proteins, thus streaking and smearing should be avoided. Protein patterns should be reproducible from gel to gel. Therefore, a crucial step performed before electrophoresis is sample preparation. However, the main issue with the plant tissues is that it contains low cellular proteins, and more content of proteases and interfering compounds such as phenolics, pigments, lipids, and nucleic acids. After extraction, proteins have to be solubilized in a solution compatible with isoelectric focusing (IEF). These are then denatured and precipitated in a mixture of β-mercaptoethanol and trichloracetic acid (TCA) in cold acetone, which helps in the efficient inhibition of protease activity in plant tissues. The proteins are then solubilized in the urea-K_2CO_3 solution, which results in highly reproducible gels with a good spot resolution in large pH and Mr ranges. The solubilization solution was developed for the first dimension in IPG strips. However, running protein samples on IPGs prompted modifications in solubilization procedures, because of the high salt content and the presence of ionic detergent (SDS) that are not compatible with the high voltages required to perform separation in such gels.

Before you begin

Preparation of gel bed.

Advanced Methods in Molecular Biology and Biotechnology. https://doi.org/10.1016/B978-0-12-824449-4.00015-3

Key resources table

Reagent or resource	Source	Identifier
TCA ß-ME (v/v) in cold acetone	Sigma-Aldrich	MFCD00004890
Urea	Sigma-Aldrich	MFCD00008022
SDS	Sigma-Aldrich	11,667,289,001
• K$_2$CO$_3$	Sigma-Aldrich	243,558
Triton X-100	Sigma-Aldrich	MFCD00128254
Urea	Sigma-Aldrich	MFCD00008022
Thio urea	Sigma-Aldrich	MFCD00008067
CHAPS	Sigma-Aldrich	850500P
Phosphine	Sigma-Aldrich	MFCD00079654
Pharmalyte	Sigma-Aldrich	GE13-0451-01

Materials and equipment

Materials and equipment required:
- Electrophoresis apparatus
- Electrophoretic tank
- Casting tray
- Comb
- Micropipettes
- Conical flasks for gel preparation
- Spectrophotometer
- Spatula

Reagents required:
- Precipitation solution: 10% TCA (w:v), 0.07% 2ME (v:v) in cold acetone. This solution must be prepared freshly and stored at $-20°C$ until use. All three components are toxic and the solution must be prepared under a hood.
- Rinsing solution: 0.07% 2ME (v:v) in cold acetone. This solution is then stored at $-20°C$ for a month.
- R2D2 solubilization solution: Urea (5 M), thiourea (2 M), CHAPS 2% (w:v), SB3–10 (2%) (W:v), DTT (20 mM), phosphine (5 mM), pharmalyte (0.5%) 4–6.5 (v:v), pharmalyte 3–10 (0.25%) (v:v) in ddH$_2$O. The solution is heated (below 30°) to help urea solubilize. It is aliquoted and then stored at $-80°C$ for months
- UKS solubilization solution:urea (9.5 M), K$_2$CO$_3$ (5 mM), SDS (1.25%), DTT (5%), Triton X-100 (6%), ampholines (2%) 3.5–9.5 in ddH$_2$O. K$_2$CO$_3$ is prepared as a 2.8% (w:v) stock solution, SDS as a 10% (w:v) filtered stock solution, and Triton X-100 is provided as a 20% solution. Besides 3 mL ddH$_2$O

are added to other components till the urea is solubilized (heating below 30 °C). A 40 mL solution is usually prepared, then aliquoted and stored at $-80°C$ for months.

- IPG strip rehydration solution: Solution A: 7 M urea, 2 M thiourea, 1.4% CHAPS (w:v), 16 mM DTT, 5 mM phosphine, 0.3% pharmalyte 3–10 (v:v) in ddH$_2$O. Solution B: 7 M urea, 2 M thiourea, 0.5% CHAPS (w:v), 10 mM DTT, 5 mM phosphine in ddH$_2$O.

Step-by-step method details

1. Protein precipitation and denaturation

- Plant tissues are ground in mortar with a pestle in liquid nitrogen and a fine powder is thus obtained.

  Note: A very fine powder should be obtained; it may be obtained by the precrushing of hard materials. An automatic cryogenic crusher with a metallic ball may be used to conduct this process.

- About 200 µL of powder is transferred to a weighted 2-mL eppendorf tube, and covered with 1.8 mL of the cold TCA-2ME-acetone solution, mixed, and then stored at $-20°C$ for 1 h. TCA and acetone have the capacity to denature and precipitate the proteins. This solution inactivates the phenoloxidases and oxidases, preventing phenol oxidation into quinones, which would result in protein binding to insoluble complexes. It was also shown to inactivate proteases, as well as phenol extraction. Acetone alone allows the solubilization of the pigments, lipids, and terpenoids possibly present in the tissue. 2ME prevents the formation of disulfide bonds during precipitation.

  Note: It is important to work the procedure at a low temperature to limit protease action.

- Centrifuge for 10 min at 10,000 × g (refrigerated centrifuge, below 4°C).

2. Rinsing with 2ME-acetone solution

- The supernatant is then discarded and the pellet is resuspended in 1.8 mL of cold rinsing solution. It should be stored at $-20°C$ for 1 h. This step is critical in the sense that it can eliminate the acidity due to TCA that would impede protein recovery.
- Centrifuge 10,000 × g for 15 min at a low temperature of 4°C.
- The supernatant is then discarded and the step is then repeated twice.
- The pellet is dried under vacuum for about 1 h to eliminate acetone.
- Weigh the pellet.

3. Protein solubilization

- The amount of UKS or R2D2 buffers for protein solubilization depends on the plant tissue; for instance, we use 60-µL/mg dry powder for leaf (maize, rape) and 50-µL/mg dry powder for maize kernel.

- *Re*-solubilization is achieved by vortexing for 1 min.
- Centrifuge at $10,000 \times g$ (at 25°C) for 15 min and collect the supernatant in a 1.5-mL eppendorf tube.
- Centrifuge again (15 min, 25°C), and the supernatant is then transferred to a new eppendorf tube. Samples containing solubilized proteins can then be stored at -50°C or -80°C for months.

4. **Preparation of samples for IEF**
   - Before using the protein samples, they must be centrifuged again. The pellet is then discarded.
   - Around 50 μg of total proteins are loaded per IEF gel (24 cm long, either classical rod gels or IPGs) for analytical gels, which are then revealed with silver nitrate, or 150–500 μg for gels which are revealed with colloidal Coomassie blue.
   - Samples solubilized in R2D2 are suitable for use in IEF with IPGs. The required volume of solubilized samples should be complemented to reach the desired protein amount with the R2D2 solubilization buffer, up to 450 μL for active rehydration.

Expected results

Several protein spots *with their compatible gels in different pH ranges can be observed after electrophoresis.*

Advantages

TCA-acetone 2D gel electrophoresis is a simple step of extraction. It is less expensive and less time-consuming. The technique produces more unique spot detection and detects high *Mr* proteins.

Limitations

This technique usually has a low protein yield.

Optimization and troubleshooting
Problem

There is usually distortion of the 2D pattern. Horizontal streaking or incomplete focused spots can be observed. Point streaking can also be observed.

Potential solution

It is important to make sure that there is no seepage during gel casting. The sample must be thoroughly and stably solubilized. Recurrent re-solubilization cycles give rise to or increase horizontal streaking. Wash glass plates appropriately. Remove any excess or residual thiol reducing agent with iodoacetamide prior to loading the IPG strips onto the second-dimension gel to avoid point streaking.

Safety considerations and standards

- The precipitation and rinsing solutions must be cold when used. It is important to keep an acetone bottle at 4°C temperature to prepare these solutions.
- Rehydration buffer contains high molarity of urea, so it must not be heated above 30°C as isocyanate ions are produced, which result in protein carbamylation.
- For efficient protein extraction, it is important to obtain it in fine powder. It can be achieved by the precrushing of hard material (e.g., mature maize grains). An automatic cryogenic crusher with a metallic ball (6 mm diameter) can also be used for crushing.
- To recover a white pellet, various rinsing steps can be applied with highly pigmented samples. It is feasible to extend rinsing overnight.
- The dry pellet powder can also be stored at − 80°C before protein solubilization. However, in this case, it is better to redry the powder prior to protein re-solubilization.
- It is important to handle all the reagents involved with gentle care.

Further reading

Ausubel, F. M., Brent, R., Kingston, R. E., Moore, D. D., Seidmean, J. G., Smith, J. A., et al. (1991). *Current protocols in molecular biology. Vol. 1.* New York, NY: Wiley Interscience (Google Scholar).

Carpentier, S. C., Witters, E., Laukens, K., Deckers, P., Swennen, R., & Panis, B. (2005). Preparation of protein extracts from recalcitrant plant tissues: An evaluation of different methods for two-dimensional gel electrophoresis analysis. *Proteomics, 5*(10), 2497–2507.

Damerval, C., Zivy, M., Granier, F., & de Vienne, D. (1988). *Two-dimensional electrophoresis in plant biology.* Laboratoire de génétique des systèmes végétaux.

Damerval, C., De Vienne, D., Zivy, M., & Thiellement, H. (1986). Technical improvements in two-dimensional electrophoresis increase the level of genetic variation detected in wheat-seedling proteins. *Electrophoresis, 7*(1), 52–54.

Granier, F. (1988). Extraction of plant proteins for two-dimensional electrophoresis. *Electrophoresis, 9*(11), 712–718.

Hames, B. D. (Ed.). (1998). *Gel electrophoresis of proteins: A practical approach* OUP Oxford. Vol. 197.

Dansyl-Edman method of peptide sequencing

16

Definition

The Edman degradation is a chemical reaction that is executed in a series of steps and removes N-terminal amino acids in a sequential manner from a peptide or protein. The overall reaction sequence is shown in Fig. 11. Coupling reaction is the first step in the technique which involves the reaction of phenylisothiocyanate (PITC) with the N-terminal amino group of the peptide or protein. This sample is later dried, followed by treatment with an anhydrous acid (e.g., trifluoracetic acid). This treatment results in cleavage of the peptide bond between the two amino acids. The PTH amino acid obtained later is identified, normally by reverse-phase HPLC. This process is called direct Edman degradation. The dansyl-Edman method for peptide sequencing has the same procedure with certain modifications such as the thiazolinone is extracted instead of being converted to PTH derivative, it is discarded. Instead, a small fraction (5%) of the remaining peptide is taken and the newly liberated N-terminal amino acid is determined in this sample by the dansyl-Edman method. Although the dansyl-Edman method results in successively less peptide being present at each cycle of the Edman degradation, this loss of material is compensated for by the considerable sensitivity of the dansyl-Edman method for identifying N-terminal amino acids.

Key resources table

Reagent or resource	Source	Identifier
0.5 M Tris HCL	Sigma-Aldrich	MFCD00012590
10% SDS	Sigma-Aldrich	MFCD00036175
Acrylamide/bis-acrylamide	Wiley/HIMEDIA	386,284,152/MB005
Ammonium per sulfate (APS)	Sigma-Aldrich	MFCD00003390
Isopropanol	Sigma-Aldrich	MFCD00011674
• SDS-sample loading buffer (Lammeli Buffer 2×, pH=6.8)	Sigma-Aldrich	S3401
SDS-sample running buffer (Tris Gly, pH=8.3)	Thermo Fisher scientific	LC2675
Coomassie staining solution	HIMEDIA	ML046
Destaining solution	Thermo Fisher scientific	P1304MP
Ladder	HIMEDIA	MBT051

Advanced Methods in Molecular Biology and Biotechnology. https://doi.org/10.1016/B978-0-12-824449-4.00016-5

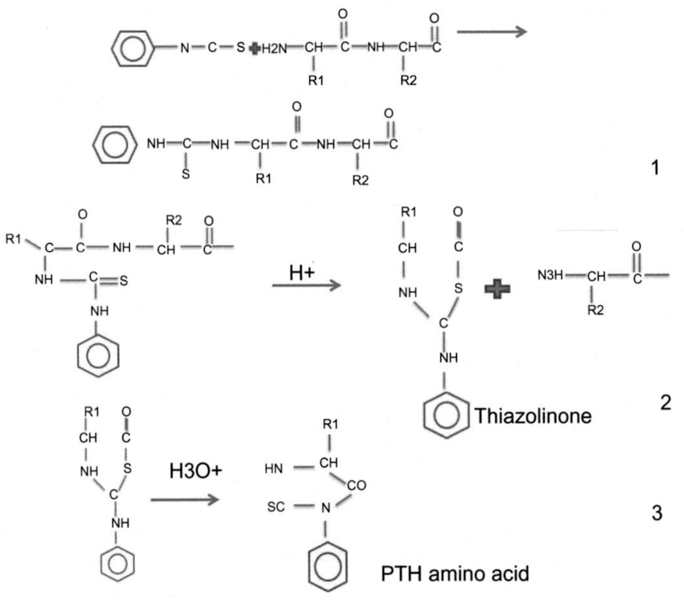

FIG. 11

The Edman degradation reactions and conversion step: (1) the coupling reaction; (2) the cleavage reaction; and (3) the conversion step.

Materials and equipment

Materials and equipment required:

- Ground glass stoppered test tubes (approximately 65×10 mm, e.g., Quickfit MF 24/0). All reactions are carried out in this "sequencing" tube.
- Pyridine (50%) (aqueous made with AR pyridine). This is stored under nitrogen at 4°C in the dark. However, some discoloration may occur after some time, but it does not affect the results of the experiment.
- 5% v/v Phenylisothiocyanate in pyridine (AR). These are stored in nitrogen at 4°C under dark conditions. The phenylisothiocyanate should be highly pure and is best purchased as "sequenator grade." Fresh stock should be made once in a month.
- Water-saturated butyl acetate should be stored at room temperature.
- Anhydrous trifluoroacetic acid (TFA). Stored at room temperature in nitrogen.

Step-by-step method details

1. The peptide to be sequenced should be dissolved in the proper volume of water, transferred to the sequencing tube, and dried to leave the film of peptide at the bottom.

2. The peptide is then dissolved in 50% pyridine (200 p-L), aliquot removed for N-terminal analysis by the dansyl-Edman method.
3. Phenylisothiocyanate (100 nL) of 5% is added to the sequencing tube and mixed. The contents are then flushed with nitrogen and incubated at 50°C for 45 min.
4. The tube is then placed in vacuo for 30–40 min.

Note: The desiccator should contain a beaker of phosphorus pentoxide to act as a drying agent, and, if possible, the desiccator should be placed in a water bath at 50–60°C.

5. A white crust is visible at the bottom of the tube after drying, which completes the coupling reaction.
6. TFA (200 IJL) is added to the test tube, flushed with nitrogen, and incubated at 50°C for 15 min.
7. After incubation, the test tubes are placed in vacuo for 5 min. TFA is a very volatile acid and evaporates rapidly. The cleavage reaction is complete with this step.
8. The contents are then dissolved in 200 mL of water.

Note: The contents of the tube do not appear to be completely dissolved as many of the side products produced in the previous reactions are not soluble.

9. Butyl acetate (1.5 mL) is then added to the tube, vigorously mixed for 10 s, and then centrifuged in a bench centrifuge for 3 min.
10. The upper organic layer is carefully discarded, taking care not to disturb the lower aqueous layer
11. The butyl acetate extraction procedure is repeated again and then the test tube containing the aqueous layer is placed in vacuo (with the desiccator standing in a 60°C water bath if possible) until dry (30–40 mm).
12. The dried material is redissolved in the test tube in 50% pyridine (200^L) and the aliquot is removed to determine the newly liberated N-terminal amino acid by the Dansyl-Edman method.
13. A further cycle of the Edman degradation can now be carried out by returning to step 3. These steps are further carried out until the peptide has been completely sequenced.

Advantages

There is a guarantee of the N-terminal sequence of proteins. No pretreatment of the sample is required. Isobaric amino acids are also differentiated, e.g., leucine and iso-leucine. Identification of the unknown proteins that have not been registered in the databases is also possible using this method.

Limitations

The technique lacks high-throughput capabilities as sequencing proceeds on samples of single proteins only.

Further reading

Edman, P., Högfeldt, E., & Sillén, L. G. (1950). Method for determination of the amino acid sequence in peptides. *Acta Chemica Scandinavica, 4*(7), 283–293.

Hartley, B. S. (1970). Strategy and tactics in protein chemistry. *Biochemical Journal, 119*(5), 805–822.

Offord, R. E. (1966). Electrophoretic mobilities of peptides on paper and their use in the determination of amide groups. *Nature, 211*(5049), 591–593.

Slater, R. J. (1986). *Experiments in molecular biology* (pp. 121–129). Totowa, NJ: Humana Press.

Walker, J. M. (1997). The Dansyl-Edman method for manual peptide sequencing. In *Protein sequencing protocols* (pp. 183–187). Totowa, NJ: Humana Press.

ISSR analysis

17

Definition

ISSR (intersimple sequence repeat) analysis is the tool for studying biodiversity and screening of the germplasms. ISSRs are DNA fragments of about 100–3000 bp that are situated between adjacent, oppositely oriented microsatellite regions. PCR is used to amplify the ISSRs using microsatellite core sequences as primers with a few selective nucleotides as anchors into the non-repeat adjacent regions (16–18 bp). The fragments from multiple loci are generated simultaneously separated by gel electrophoresis and scored accordingly. There are certain techniques related to ISSR analysis and those are single primer amplification reaction (SPAR) and directed amplification of Minisatellite-region DNA (DAMD). The technique of ISSR analysis has been also used for DNA fingerprinting and determining the genetic diversity in germplasm. ISSR markers are noncoding loci and are disseminated all through the genome. ISSR markers are known to have high reproducibility, can produce a large number of fragments, and are also cost-effective, which makes it a potential tool for evaluating changes in diversity in agronomically and commercially important crops as well as genetically diverse plant and animal species. Here we devise the protocol for ISSR marker identification in relation to genetic divergence studies.

Before you begin

- Prepare ISSR primers to the concentration of 10 nmol by diluting 10 μL of ISSR primer (stock solution) in 90 μL dH$_2$O.
- Preparation of the gel bed.

Key resources table

Reagent or resource	Source	Identifier
• BSA5,	Sigma-Aldrich	MFCD00130384
• ISSR primers	Thermo Fisher scientific	–
• 1 × TBE buffer	Sigma-Aldrich	574,795
• Agarose	Sigma-Aldrich	Mfcd00081294
• DNA ladder	Sigma-Aldrich	D0428

Advanced Methods in Molecular Biology and Biotechnology. https://doi.org/10.1016/B978-0-12-824449-4.00017-7

Materials and equipment

Materials and equipment required:
- Gloves (sterile)
- Microtubes (autoclaved)
- Pipettes
- Thermocycler
- Analytic scales
- Erlenmeyer flasks
- Sterile tips for micropipettes
- Electrophoresis apparatus
- Microwave oven
- Transilluminator

Reagents required:
- DNA (2–5 ng/μL),
- H_2O (sterile, nuclease-free, distilled),
- BSA5,
- ISSR primers
- Agarose,
- 1 × TBE buffer
- DNA ladder (100 bp)

Step-by-step method details

1) **PCR (final volume 20 μL/one reaction)**
   a) The reagents as well as DNA sample need to be thawed at room temperature (or by hand or in a thermoblock).
   b) All vials are centrifuged a little before use and each microtube is labeled according to the sample.
   c) For more than one sample, the premix is prepared separately that contains no DNA to a volume respective to the number of the samples.

   One single reaction possesses:

   7.3 μL H_2O
   10 μL MasterMix (2 × concentrated)
   0.5-μL selected primer
   0.2 μL BSA
   d) The premix is vortexed and centrifuged briefly followed by the division of 18 μL of the premix into each microtube (0.2 mL).
   e) 2 μL DNA is added into the microtubes, shortly vortexed, and centrifuged (to get rid of air bubbles).
   f) These are then put immediately into the thermocycler and before choosing the right programs for ISSR primers:

g)
1. Denaturation 94°C/5 min
2. 40× repetitions 94°C/1 min specific temperature for the chosen primer/1 min 72°C/2 min
3. Final extension 72°C/10 min
4. Storage 4°C/forever
1. **Electrophoresis**
   a) The first well is left empty and 5 μL of the samples are put in (product of PCR)

Note: While pipetting the sample, care should be taken as there is a high risk of perforating the gel by the bottom of the tip.

b) 2 μL of the ladder is put into the first and the last well.
c) The bath is covered with the gel and buffer at the positions of the wires attached.
d) The source of the current should be set to set to 55 V and 110 mA. When the system is closed and secured, the electricity source may be plugged in and turned on.
e) Bubbles that are raised from the wires show that the electrophoresis is working fine. It can be allowed to run for at least 3 h (good to control also the position of the dye added in the master mix, whose first fraction should be close to the end of the gel).
f) The source is turned off and the gel is taken out and put into the visualization machine. White light can be turned on to fix the gel on the transilluminator.
g) White light is replaced by UV light if the gel does not move. Photographs are taken.

Expected outcomes

DNA bands of different molecular weights are visible in the gel. Phenetic relationships can be interpreted for all the ISSR assays using a statistical software package.

Advantages

ISSR has a higher reproducibility than RAPDs. Bands are homologous (preliminary results) and hypervariable enough for within-population.

Limitations

As ISSR is a multilocus technique, there is a possibility of non-homology of similar sized fragments.

Safety considerations and standards

- Ensure there is no leakage in the gel caster plate before you add the resolving gel.
- Ensure no bubbles are left on the face of the stacking gel and in the wells. This will ensure better protein resolution.
- Do not forget to wear standard gloves while handling the reagents and Gel.
- Before loading, keep all the samples on ice.
- Avoid heating the sample for more than 5 min.
- To avoid overstaining that could possibly delay results, do not stain the gel for more than 5 min.

Alternative methods/procedures

- RAPD analysis
- SSR analysis
- AFLP analysis

Further reading

Godwin, I. D., Aitken, E. A., & Smith, L. W. (1997). Application of inter simple sequence repeat (ISSR) markers to plant genetics. *Electrophoresis*, *18*(9), 1524–1528.

Sarwat, M., Nabi, G., Das, S., & Srivastava, P. S. (2012). Molecular markers in medicinal plant biotechnology: Past and present. *Critical Reviews in Biotechnology*, *32*(1), 74–92.

Sarwat, M. (2012). ISSR: A reliable and cost-effective technique for detection of DNA polymorphism. In *Plant DNA fingerprinting and barcoding* (pp. 103–121). Humana Press.

Sucher, N. J., Hennell, J. R., & Carles, M. C. (2012). *Plant DNA fingerprinting and barcoding: Methods and protocols*. New York, USA: Humana Press.

Gene cloning/recombinant DNA technology

18

Definition

The process of making multiple exact/identical copies of a specific gene or piece of DNA (clones) using the techniques of genetic engineering is known as gene cloning/ DNA cloning/molecular cloning/recombinant DNA technology. This method is particularly used for gene isolation and its amplification.

Rationale

A fragment of DNA, which contains the gene of interest, is inserted into a suitable vector/plasmid in order to produce a recombinant DNA molecule. The gene is transported via a vector into the host cell, which acts as a vehicle. This vector then replicates, producing a number of identical copies of itself as well as of the gene that carries it. After the division of the host cells, the copies of the recombinant DNA molecule are passed to the next generation and further vector replication takes place. A large number of cell divisions result in the production of a clone or a group of identical host cells. Each cell in the clone contains one or more copies of the gene of interest.

Materials and equipments

- **Host**—in which the gene of interest is inserted/recombinant DNA replicates and it should be viable, e.g., E*scherichia coli*-DH5α
- **Vector**—to carry, maintain, and replicate the gene of interest in the host cell, e.g., bacteriums, plasmids, bacteriophages, bacterial artificial chromosomes (BACs), mammalian artificial chromosomes (MACs), etc., e.g., pEGFPC1.
- **Gene of Interest**—the desired genes to be cloned.

Attributes of an ideal vector

An ideal vector should possess the following:

- **Marker for selection**—which helps to distinguish the vector containing cells from other cells, e.g., Antibiotic resistance gene.

Advanced Methods in Molecular Biology and Biotechnology. https://doi.org/10.1016/B978-0-12-824449-4.00018-9

- **Origin of replication**—a point at which DNA replication starts, which facilitates DNA opening, i.e., self-replicating nature. This site is usually AT rich and contains restriction and ligase enzymes.
- **Small size**. E.g., 4.7 kb.
- **Multiple/poly cloning site (MCS/PCS)**—a DNA region within a plasmid containing multiple, specific, distinctive restriction enzymes cut sites or recognition sequences, where a number of restriction enzymes can cut or cleave after recognizing their recognition sites.
- **High copy number**—to produce more copies of the gene of interest, e.g., recombinant strain of *E. coli* DH5α contains a high copy number of plasmids due to the presence of a mutated RoP gene.
- **Marker for recombination**—which helps in Blue-White screening/selection of clones, e.g., lac Z gene in *E. coli*.
- Introduction of donor DNA fragment must not interfere with the replication property of the vector.
- Its isolation from the host cell should be easy.

Attributes of a good host cell

- It should permit the comfortable entry of the recombinant DNA into the cell.
- Should not possess endogenous DNA deteriorating enzymes and shall not destroy the recombinant DNA it uptakes.
- Should be able to stably maintain and contain the recombinant DNA.
- The transformed host must not independently sustain outside the laboratory.
- Should be easy to maintain and handle.
- Should be available in a wide range of genetically defined strains.
- Should accept a large cohort of different vectors.

Step-by-step method details

- Isolation of Vector from *E. coli*.
- Isolation of gene of interest to be cloned. DNA in case of prokaryotes and mRNA in case of Eukaryotes.
- Primer designing for amplification of gene of interest. The primers should have restricting sites for enzymes that are in MCS of vector. Selection of Restriction sites on the primer should be such that these restriction sites should NOT be anywhere in the gene part to be amplified.
- If cloning is for expression, primers shall be such that the start signal of the gene sequence should be in frame with the vector.
- PCR amplification of gene of interest.
- Gel elution of gene of Interest after PCR amplification.
- Restriction digestion of vector and gene of interest.

- A second gel elution of restricted gene of interest and restricted vector.
- Ligation of vector and insert.
- Transformation of recombinant DNA molecule into a suitable host cell.
- Selection of host cells to be transformed and identification of the clone which contains the gene of interest.
- Multiplication/expression of the introduced gene in the host.
- Isolation of multiple gene copies or protein expressed by the gene.
- Purification of the isolated gene copy or protein.

In the detailed example below, we have described the cloning of a CBF gene from Tomato in the pEGFPC1 vector to make a fusion protein of GFP-CBF1.

Plant material

Tomato seeds were collected from seed repositories of SKUAST-Kashmir. The seed material was soaked in water for 10 min, washed twice with sterile water, and then sterilized seeds were germinated in petriplates at 24°C in the incubator. Germinated seedlings were transferred to petriplates and grown under controlled conditions in a growth chamber at 24°C/15°C (Day/Night) and were processed for RNA extraction.

Microbial cultures and plasmids

E. coli (DH5α) was obtained from the National Institute of Immunology, New Delhi, India. The pEGFPC1 was kindly supplied by Prof Zhou Wang, University of Pittsburgh, Pennsylvania, USA.

DEPC treatment

Microfuge tubes, tips, tip boxes, mortar and pestle, and spatula were DEPC treated overnight and then autoclaved followed by drying in a hot air oven.

RNA extraction and qRT-PCR

Total RNA was extracted at the germination stage from the seedlings of tomato using the Trizol (Invitrogen) (Chomczynski & Sacchi, 1987) method as follows.

Tissue samples (100 mg) were ground in a mortar and pestle in liquid nitrogen and transferred to a new microfuge tube. To it, 1 mL of Trizol was added to the homogenized sample and mixed thoroughly and incubated at room temperature for 5 min; 0.2 mL of chloroform was added and the tubes were shaken vigorously for 15 s. The tubes were then incubated at room temperature for 2–3 min. The samples were centrifuged at $12,000 \times g$ for 15 min at 4°C. The mixture was then separated into lower red, an interphase, and a colorless upper aqueous phase. The aqueous phase exclusively contains RNA. The upper aqueous phase was carefully transferred into a new 1.5 mL tube without disturbing the interphase. The aqueous phase was processed for RNA precipitation. To this aqueous phase 0.5 mL of 100% isopropanol was added. Incubation at room temperature for 10 min was done. Centrifugation was done at $12,000 \times g$ for 10 min. The supernatant was removed from the tube with the RNA pellet remaining inside. The pellet was then washed with 1 mL of 75% ethanol. The samples were vortexed briefly and then the tubes were centrifuged at $7500 \times g$ for 5 min at 4°C. The RNA pellet was then air-dried for 5–10 min and it was resuspended in RNase free water (DEPC water). After dissolving, samples were run on a 1% agarose gel to analyze the quality of RNA and the presence of any DNA contamination.

DNase treatment of RNA samples

The Thermo Scientific DNase kit was used for removing traces of DNA in the extracted RNA. The DNase treatment was given following the manufacturer's protocol.

Quantitative and qualitative analysis of RNA

The quality of RNA was determined by using Nanodrop (*Thermo Fisher SCIENTIFIC*) at an OD of 260 and 280 nm. The samples showing an $OD_{260/280}$ ratio between 1.9 and 2 were used for further experimental studies. RNA quantity was estimated by using an OD of 260 nm.

cDNA synthesis

cDNA synthesis was done using the Thermo Scientific Revert Aid First Strand cDNA Synthesis Kit using oligo dT primers. The first strand cDNA was synthesized by using the manufacturer's protocol and then stored at -20°C. The protocol for the synthesis of first strand is given below.

The following reagents were added in a tube free from the RNase free tube on ice as per the given order.

Template	Total RNA	1.5 µg
Primers	Oligo (dT)$_{18}$ primer	1 µL
Nuclease free water		Final vol to 12 µL
Total volume		**15 µL**

The components were gently mixed and a brief spin was given. Incubation at 65°C for 5 min was given. The tube was chilled on ice and a brief spin was given. The tube was again placed on ice and the following components were added as per the indicated order.

5× reaction buffer	4 µL
RiboLock RNase inhibitor (20 U/µL)	1 µL
10-mM dNTP mix	2 µL
RevertAid M-MuLV RT (200 U/µL)	1 µL

The components were gently mixed and a brief spin was given. The components were incubated at 42°C for 60 min. The reaction was terminated by heating at 70°C for 5 min.

Primer designing

For designing primers for amplification of genes coding for CBF1 from Solanum lycopersicon, amino acid sequences for the previously submitted Genbank sequence were retrieved from GenBank (Farheena, 2018—MSc Thesis). The mRNA sequence was aligned using Clustal W (www.ebi.ac.uk/clustalw) to check for conserved motifs (Chenna, Sugawara, Koike, et al., 2003; Larkin et al., 2007). Two oligonucleotide primers for the CBF1 gene were designed using the primer 3 online software. A restriction site was added in the primers at the 5′ and the 3′ ends (see table further).

Validation of cDNA and primers using PCR

Polymerase chain reaction using primers of the GAPDH gene was carried out to validate the cDNA synthesis and primers. Details of the primers used are given in the following table.

Primers for amplification of CBF-1 and GAPDH genes. The bold underlined sequence represents the site for the restriction enzyme.

Nature of primer		Primer sequence	Size of amplicon	T_m (°C)	Restriction site
GAPDH	Forward primer	5′ CCTTAGTTGATT CTGTACC3′	169 bp	59.7	Nil
	Reverse primer	5′ GCCAGTTAGTAGT AATTGCCA 3′		58.3	Nil
CBF1	K-Lab-CBF1-LS-F	CTCAGT**AGATCT**ATGAATAT CTTTGAAACCTA	618 bp	55	*Bgl*ll
	K-Lab-CBF1-LS-R	CGTACT**CTGCAG**TTAGATA GAATAATTCCATA		55	*Pst*l

PCR reaction mixture

PCR amplification for CBF-1 and GAPDH genes was carried out in a reaction volume of 20 μL in 0.2 mL PCR tubes. The reaction mixture contents are given in the table below.

PCR reaction mixture for amplification of CBF-1 and GAPDH genes.

Constituents	Volume
Autoclaved water	13.6 μL
PCR buffer (1×)	2 μL
$MgCl_2$ (25 mM)	1.2 μL
dNTPs (25 mM)	0.4 μL
Forward primer (10 μM)	1 μL
Reverse primer (10 μM)	1 μL
TaqPolymerase (5 U/μL)	0.3 μL
cDNA sample (70 ng/μL)	0.5 μL
Total volume	**20 μL**

PCR cyclic conditions

The amplification reaction was carried out in a thermocycler (*Applied Biosystems*). The cycling conditions are shown in the following table.

Thermal cycling conditions for amplification of CBF-1 and GAPDH genes.

Steps		Temperature (°C)	Time	Number of cycles
1. Hot-Start		94	2 min	1
2. Denaturation		98	10 s	35
3.Annealing	GAPDH	61	30 s	
4. Extension		68	2 min	
5. Final extension		68	10 min	1

Gel electrophoresis

The amplified products of GAPDH and CBF1 genes were electrophoresed on a 2% agarose gel and compared with the 100 bp DNA ladder respectively.

cDNA quantification

cDNA quantification was done using a Qubit$^{@}$ 2.0 Fluorometer (Invitrogen).

Confirmation of cDNA synthesis

The internal gene primers for GAPDH used for confirmation of cDNA synthesis were designed (Table 1) using the protocol outlined in the primer3 online software (http://frodo.wi.mit.edu/primer3/) (Rozen & Skaletsky, 2000). The primers were checked for folds and self-complementarity by oligocalc software (http://www.basic.northwestern.edu/biotools/oligocalc.html) (Kibbe, 2007). These GAPDH primers were used for confirmation of cDNA synthesis. The predicted size of the amplicon was 169 bp.

Table 1 Primers for amplification of CBF1 and GAPDH genes.

Nature of primer		Primer sequence	Size of amplicon	T$_m$ (°C)
GAPDH	Forward primer	5′ CCTTAGTTGATTCTGTACC 3′	207bp	59.7
	Reverse primer	5′ GCCAGTTAGTAGTAATTGCCA 3′		58.3

Gene cloning

CBF1 was cloned from cDNA of *Solanum lycopersicon* Mill. The strategy from cloning of CBF-1 in pEGFPC-1 is depicted in Fig. 12.

PCR amplification of full-length genes (CBF-1)

Total RNA was extracted from 1 to 2 week-old seedlings of *Solanum lycopersicum* Mill. Var Pusa Sheetal variety. First strand cDNA was prepared with 2 µg of total RNA.

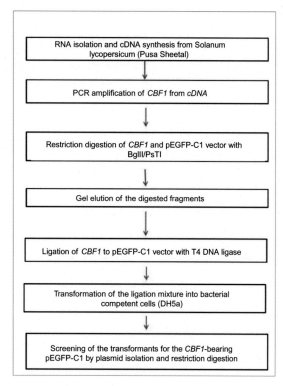

Strategy from cloning of CBF-1 in pEGFPC-1

FIG. 12

Strategy from cloning of CBF-1 in pEGFPC-1.

The putative full-length *CBF1*cDNAs were amplified with K-Lab-CBF1-LS-F and K-Lab-CBF1-LS-R primers. After an initial denaturation at 94°C for 2 min, 35 cycles were run each with 30 s of denaturation at 94°C, followed by 40 s annealing at 55°C and a 1 min extension at 72°C.

The following constituents were mixed gently and the tubes were spin briefly:

Constituents	CBF1
Buffer (10×)	2.5 μL
dNTP's (2.5 mM each)	2.5 μL
MgCl$_2$ (25 mM)	1.0 μL
cDNA (PB)	1.0 μL
Vent DNA Pol.	1 U
Gene specific primers F+R (10 mM)	2+2 μL

The PCR reaction was run in an Eppendorf thermocycler as per the following protocol.

The protocol followed for the PCR is as follows:

S. no.	Gene	CBF1	
	Step	Temperature (°C)	Time (min)
1.	Initial denaturation	95	3:00
2.	Denaturation	95	0:30
3.	Tm	50	0:30
4.	Elongation	72	1:30
5.	Step 2 repeat 30 cycles		
6.	Final extension	72	10:00
7.	Hold	4	∞

The products were run on a 1% agarose TAE gel and visualized on a UV-transilluminator. Photographs were taken using a gel documentation system. A 1 kb/100 bp ladder (Genetix) was used as a molecular weight marker. The DNA bands were excised from the gel with a sharp sterilized blade, weighed, and put in an autoclaved 2 mL microfuge tube. DNA purification from the excised band was carried out by using the MinElute gel purification kit (QIAGEN) according to the manufacturer's instructions.

Restriction digestion of the PCR fragments and vector

The eluted fragments of CBF1 and pEGFPC1 vector were double digested simultaneously with the specific restriction enzymes using the following protocol.

Constituents	CBF1 (μL)	pEGFPC1
DNA	5	5 μL
Buffer (5×)	2	2 μL
Restriction enzyme (thermo) (1 U)	BglII and *PstI*	BglII and PstI
H₂O	1	1 μL

The above constituents were gently mixed and the tubes were spun briefly and incubated at 37°C for 1 h. The products were run on a 1% agarose TAE gel and visualized on a UV-transilluminator and photographed using a gel documentation system. A 1 kb ladder was used as a molecular weight marker. The restricted fragments were gel purified using the MinElute gel purification kit (QIAGEN) according to the manufacturer's instructions.

Expression vector

The expression vector used was pEGFPC1 obtained from Zhou Wang, University of Pittsburgh, PA, USA. Vector preparation was carried out by double digestion of pEGFPC1 expression vectors using restriction enzymes as given in the Table above. A schematic diagram for cloning of the CBF-1 gene in the pEGFPC-1 vector is depicted in Fig. 3.

Ligation

The ligation reaction was performed using 0.2 μg of the digested vector, 0.4 μg insert, 1 μL ligation buffer, and 1 U T4 DNA ligase in a final volume of 10 μL. The samples were incubated at 4°C for 24 h and the recombined plasmid was transformed into *E. coli* DH5α and BL21 (DE3).

Transformation

A single colony of *E. coli* DH5α was inoculated into 10 mL LB broth in a 50 mL falcon tube. It was grown at 37°C for 16 h. 1 mL of the above grown culture was used to inoculate 100 mL of LB in a 250 mL flask and kept in an incubator shaker at 37°C for 4 h with constant shaking (250 rpm). The culture was kept on ice for 10 min and transferred to a 50 mL falcon tube and centrifuged at 5000 rpm for 5 min at 4°C. After centrifugation, the supernatant was decanted and the cells were resuspended in 1 mL cold 0.1 M CaCl₂. The cells were vortexed and again centrifuged at 5000 rpm for 5 min at 4°C. Centrifugation was conducted and the supernatant was decanted. The cells were resuspended in 1 mL cold 0.1 M CaCl₂. The cells were vortexed and incubated on ice for 20 min to make them competent. The competent cells were dispensed in 2 mL microfuge tubes (200 μL/tube) and stored at −80°C for further use.

5 μL of the ligation mixture was pipetted gently into 200 μL of competent cells. The tubes were incubated for 30 min on ice, transferred for 2 min to 42°C (heat shock), and immediately put back on ice. 900 μL of LB was added and the tubes were incubated at 37°C for 2 h. 100 μL of the cells were plated on LB Ampicillin (50 μg/mL)/ plates. 100 μL of the control DNA was also plated on LB Ampicillin (50 μg/mL) plates. The plates were grown overnight at 37°C in an incubator.

Confirmation of cloning by restriction digestion

The clones were separately inoculated in 5 mL LB broth and incubated for 16 h with 250 rpm constant shaking at 37°C. Plasmid DNA was isolated from clones using the plasmid purification kit (Sigma) following the manufacturer's instructions. The plasmids were stored at −20°C for further use. Restriction digestion of the vectors was carried out by the following protocol:

Gene	CBF1	Empty vector
Plasmid clone (0.5 μg)	pKZ-CBF1	pEGFPC1
Buffer (1.5 μL)	Tango	Tango
R. E. – 1	BglII	BglII
R. E. – 2	PstI	PstI

The samples were incubated at 37°C for 4 h for complete digestion. Enzymes were inactivated at 65°C for 10 min. The products were run on a 0.7% agarose TAE gel at 8 V/cm for 30–45 min. The products were visualized and photographed.

Heterologous expression

The expression vector pEGFP-C1 (Clontech) was used to generate fusion protein constructs with GFP at the N terminus of CBF1 for convenient visualization using fluorescent microscopy as described earlier (Dar et al., 2014).

Cell culture experiments

Human C4-2 prostate cancer cells were obtained from Dr. Zhou Wang, University of Pittsburgh, PA, USA and maintained in RPMI1640 medium supplemented with 10% FBS, 1% glutamine, 100 units/mL penicillin, and 100 mg/mL streptomycin (Invitrogen) at 37°C in the presence of 5% CO_2 in a humidified incubator. GFP-CBF1 and GFP expression vector was transiently transfected into C4-2 cells using Lipofectamine Reagent according to the manufacturer's protocol (Thermo). Cells were transfected at >60% confluence in phenol red-free OptiMEM. The Expression of GFP-CBF1 fusion proteins was imaged 16 h after transfection using the LMI Fluorescent Microscope, ABE, London, UK.

Expected outcomes

Multiple identical copies of gene of interest.
 Functional GFP fusion protein.

Advantages

- Facilitates isolation of a particular gene and determination of its nucleotide sequence.
- Helps in identification of control sequences of DNA.
- Helps in analyzing the control sequences of DNA.
- Helps in investigating the functions of RNA/proteins/enzymes.
- Facilitates the identification of mutations. E.g., gene defects
- Facilitates the engineering of organisms for specific purposes. E.g., insect resistance, insulin production, etc.

Limitations

- Comes with a degree of uncertainty.
- Has the tendency to bring new diseases.
- Might lead to problems in organ rejection.
- Gene diversity may be decreased.

References

Chenna, R, Sugawara, H, Koike, T, et al. (2003). Multiple sequence alignment with the Clustal series of programs. *Nucleic Acids Research*, *31*(13), 3497–3500. https://doi.org/10.1093/nar/gkg500.

Chomczynski, P., & Sacchi, N. (1987). Single-step method of RNA isolation by acid guanidinium thiocyanate-phenol-chloroform extraction. *Analytical Biochemistry*, *162*(1), 156–159. https://doi.org/10.1006/abio.1987.9999.

Dar, J. A., Masoodi, K. Z., Eisermann, K., Isharwal, S., Ai, J., Pascal, L. E., et al. (2014). The N-terminal domain of the androgen receptor drives its nuclear localization in castration-resistant prostate cancer cells. *The Journal of Steroid Biochemistry and Molecular Biology*, *143*, 473–480.

Kibbe, W. A. (2007). OligoCalc: An online oligonucleotide properties calculator. *Nucleic Acids Research*, *35*(Web Server issue), W43–6.Larkin et al., 2007Larkin, M. A., Blackshields, G., Brown, N. P., Chenna, R., McGettigan, P. A., McWilliam, H., et al. (2007). Clustal W and Clustal X version 2.0. *Bioinformatics*, *23*, 2947–2948.

Rozen, S, & Skaletsky, H. (2000). Primer3 on the WWW for general users and for biologist programmers. *Methods in Molecular Biology*, *132*, 365–386. https://doi.org/10.1385/1-59259-192-2:365.

Polymerase chain reaction (PCR)

19

Definition

A polymerase chain reaction is a technique or method used to rapidly amplify the target DNA sequences. PCR has the ability to provide identical copies of a known DNA sequence; hence, it is also known as "*Molecular Phototyping*." It is a versatile technique which is used to permit the targeted amplification of a specific DNA sequence which is within the source of DNA. It enables workers to generate millions of copies of a particular gene fragment without its cloning in a shorter time period. This technique becomes immensely significant where the DNA specimen quantity is very low to work on.

In 1985, **Kary Mullis** in California, USA invented PCR that earned him the Nobel Prize in 1993. PCR is solely a biochemical in vitro method and has now been fully automated, i.e., carried out by a machine itself (Figs. 13–15).

Rationale

The principle of PCR is primarily focused on the capability of a DNA polymerase analog enzyme to produce a complementary DNA strand over an available template serving strand. The synthesis involves the primer-mediated amplification driven by enzyme action. A primer is required to direct DNA synthesis as polymerase adds nucleotides to its $3'$-OH end. It leads to the elongation of the chain of complementary nucleotides to produce a double-stranded DNA specimen. PCR is a kind of chain reaction as newly synthesized DNA strands act as templates for the next cycle and the resultant strands are made available as templates for the next cycles. With each cycle, the single segments of the double-stranded DNA template are amplified into double-stranded DNA segments. It consists of a series of cycles based on three successive steps as will be discussed later. The cycles are repeated theoretically to enable exponential increase in the DNA molecules.

Types of PCR

- Real-time PCR (R-PCR)
- Quantitative real-time PCR (Q-RT PCR)
- Reverse Transcriptase PCR (RT-PCR)
- Multiplex PCR

Advanced Methods in Molecular Biology and Biotechnology. https://doi.org/10.1016/B978-0-12-824449-4.00019-0

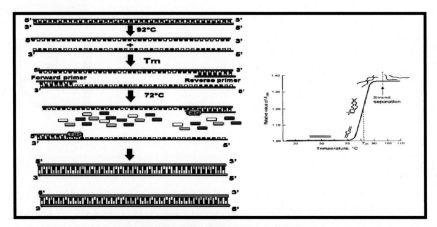

a. Polymerase Chain Reaction

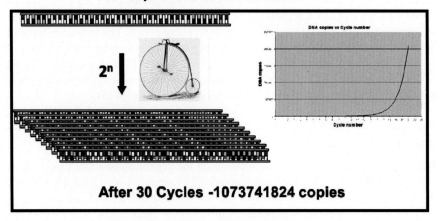

b. DNA copies after 30 cycles

FIG. 13

(A) Polymerase chain reaction. (B) DNA copies after 30 cycles.

- Nested PCR
- Long-range PCR
- Single-cell PCR
- Fast-cycling PCR
- Methylation-specific PCR (MSP)
- Hot start PCR
- High-fidelity PCR
- In situ PCR
- Variable Number of Tandem Repeats (VNTR) PCR
- Asymmetric PCR
- Repetitive sequence-based PCR

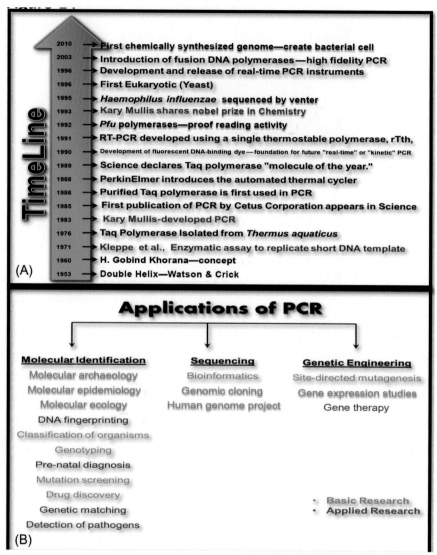

FIG. 14

(A) Timeline of developments in PCR technology. (B) Applications of PCR Technology.

- Overlap extension PCR
- Assemble PCR
- Inter-sequence-specific PCR (ISSR)
- Ligation-mediated PCR
- Methylation-specific PCR
- Mini-primer PCR

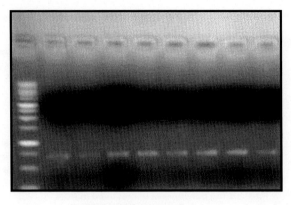

PCR gel photograph

FIG. 15

PCR gel photograph.

- Solid-phase PCR
- Touchdown PCR, etc.
- Suicide PCR

Before you begin

- Preparation of PCR reaction mixture for amplification of DNA

Constituents	Volume (μL)
Autoclaved water	18.75
PCR buffer (10×)	2.5
MgCl₂ (25 mM)	0.5
dNTPs (25 mM)	0.5
Forward primer (10 μM)	1
Reverse primer (10 μM)	1
Taq Polymerase (5 U/μL)	0.25
DNA sample (60 ng/μL)	0.5
Total volume	**25**

- Setting up thermal cycling conditions for amplification of DNA

Steps	Temperature (°C)	Time	Number of cycles
1. Initial denaturation	94	5 min	1
2. Denaturation	94	30 s	35
3. Annealing	Tm	45 s	
4. Extension	72	1 min 50 s	
5. Final extension	72	15 min	1

Key resources table

Reagent or resource	Source	Identifier
PCR buffer	Sigma-Aldrich	P2317
MgCL2	Sigma-Aldrich	Mfcd00011106
dNTPs	Sigma-Aldrich	DNTP100-1KT
Forward primer	Sigma-Aldrich	–
Reverse primer	Sigma-Aldrich	–
Taq polymerase	Sigma-Aldrich	MFCD00166026
DNA sample	–	–

Materials and equipment

- DNA template:
  It is the source of the double-stranded DNA (dsDNA) segment that is to be amplified. It serves as the master copy or template.
- DNA polymerase:
  It involves suitable thermostable DNA polymerases that could withstand temperature as high as 98°C. *Taq Polymerase is the* most commonly used polymerase, which is obtained from *Thermus aquaticus* bacteria whose natural habitat is hot springs. It has an optimum working temperature of 80°C.
- Oligonucleotide primers:
  They include two short fragments (Oligonucleotides) of single-stranded DNA made of 15–30 nucleotides which are specific for the target sequences and are complementary to the 3′ ends of the sense and antisense strands of the target sequence so that they are extended toward each other in the opposite direction.
- Deoxynucleotide triphosphates (dNTPs):
  They include single units of all DNA-forming nitrogenous bases in their deoxynucleoside triphosphates (dATP, dTTP, dGTP, and dCTP), which form the building ingredients for DNA synthesis. Besides, they are known to be important in catering to the energy demand during polymerization or chain elongation.
- Magnesium ion (Mg^{2+}):
  Magnesium ion in the +2 oxidation state acts as a cofactor and increases the enzyme activity of DNA polymerases. It catalyzes the bond formation between the 3′ end and phosphate group of a Primer and dNTP, respectively. Thus, it enhances polymerization or chain elongation during the process.
- Buffer:
  It includes the buffer that hosts the reaction and provides an ideal environment for PCR activity. The pH lies between 8.0 and 9.5, and Tris-HCl helps in its stabilization.
- Distilled water:
  To make up the final volume of the reaction.

Step-by-step method details

It consists of a series of amplification cycles of three successive steps each, i.e., Denaturation, Annealing, and Extension.

- Denaturation:
  In this step, the DNA fragment containing the desired sequence is heated for denaturation at 94°C for 15 s. Denaturation enables the separation of complementary strands by breaking the hydrogen bonds holding the DNA strands together in a helix. The separated strands act as templates for copying of DNA.
- Annealing:
  In this step, the annealing of two primers to the denatured strands takes place. The annealing temperature is comparatively low and lies between 50°C and 70°C. Low temperature enables the hybridization between templates and primers but it should not be so low (> 50°C) as to lead to mismatching.
- Extension
  In this step, the deoxygenated nucleoside triphosphates and the selected heat stable DNA polymerase are added to the reaction mixture and heated to 72°C. DNA polymerase leads to primer extension or polymerization by adding nucleotides sequentially onto the primers already bound to the desired DNA strands.

Expected outcomes

The PCR products, i.e., amplified gene products (Amplicon), are then detected by using a labeled probe specific for the desired DNA sequence. The probe produces radioactive, colorimetric, fluorometric, or chemiluminescent signals which depends on the nature of the reporter molecule used. This probe-based detection method has two purposes:

- Permits visual representation of the PCR product.
- Ensures that the target sequence is the Amplicon, which is not the result of nonspecific amplification by providing specificity.

Advantages

The technique produces millions of copies of a single DNA segment in a shorter period of time. It has a specific amplification and the procedure is rapid.

Limitations

Running and setting up requires very high technical skills. It has a high equipment cost and requires a high sterile environment.

Applications of PCR

- Identification and characterization of infectious agents (HIV, Malaria, Hepatitis, etc.)
- DNA fingerprinting (forensic application/in solving paternity disputes)
- Deciphering mutation (for diagnosis of genetic disorders)
- Gene cloning.
- DNA sequencing
- Taxonomy (for identification of species)
- Biodiversity (studying species abundance, richness, etc.)

Applications of PCR in agricultural sciences

- In identifying the mutated gene EDA in Holstein cattle with ectodermal dysplasia.
- In identifying polymorphism in the ABCB1 gene in phenobarbitol responsive-resistant idiopathic epileptica.
- In detecting the Meq gene (that decreases immune-suppression in chickens).
- In identifying the mutated gene with X-linked Myotubular myopathy in dog species.
- In identifying the Insertion mutation in ABCD4 with gall bladder mucocele formation in dogs.
- In identifying various plant and animal pathogens.
- In identification of the Bursal diseases virus (BDV) in avis.
- In identifying the Bovine respiratory syncytial virus.
- Identification of Canine Parvo-virus type-2 in dogs.
- Identification of Feline immunodeficiency virus.
- Identification of Feline leukemia virus (FeLV).

Optimization and troubleshooting
Problem
Standard PCR conditions do not yield the desired amplicon.

Potential solution
Manual optimization of PCR.

Alternative methods/procedures
- Loop-mediated isothermal amplification.
- Nucleic acid sequence-based amplification.

- Strand displacement amplification.
- Rolling circle amplification.
- Ligase chain reaction.

Further reading

Gibbs, R. A., Chamberlain, J. S., Cagan, R. L., Ready, D. F., Kidd, S., Baylies, M. K., et al. (1989). The polymerase chain reaction: A meeting report 1095. *Genes & Development, 3,* 1095–1098.

Metzker, M. L., & Caskey, T. C. (2001). Polymerase chain reaction (PCR). *eLS, 1,* 1–10. https://doi.org/10.1002/9780470015902.a0000998.pub2.

Mullis, K. B. (1990). The unusual origin of the polymerase chain reaction. *Scientific American, 262*(4), 56–65.

Mullis, K. B., & Faloona, F. A. (1989). Specific synthesis of DNA in vitro via a polymerase-catalyzed chain reaction. In *Recombinant DNA methodology* (pp. 189–204). Academic Press.

Tindall, K. R., & Kunkel, T. A. (1988). Fidelity of DNA synthesis by the *Thermus aquaticus* DNA polymerase. *Biochemistry, 27*(16), 6008–6013.

Blotting/hybridization and its types

Definition

Though a mixture of DNA/RNA or proteins can be separated by electrophoresis and the separated bands can be stained directly in the gel. However, to confirm the identity of these bands or to compare them with a known molecular probe, hybridize these bands with a labeled probe. To facilitate the hybridization of separated DNA fragments with bigger probes, the bands are transferred to a Nitrocellulose or PVDF or Nylon membrane through a technique called blotting or hybridization. These molecules (DNA/RNA or proteins) can be then visualized by using dyes like ethidium bromide, crystal violet, safranine, osmium tetroxide, etc.

Types of blotting

- Southern blotting
  The name Southern blotting is coined after its developer **E.M. Southern**. It is a blotting technique for DNA, i.e., used for analyzing DNA sequences in a DNA fragment (Fig. 16).

Step-by-step method details

It involves the following steps:

- Firstly, the DNA sample is digested with restriction enzymes producing a number of DNA fragments of unequal length.
- These fragments are subjected to gel electrophoresis, resulting in separation of DNA molecules depending on their size.
- The DNA fragments present in this gel are denatured by alkali treatment.
- The gel is laid on top of a buffer-saturated filter paper and a sheet of nitrocellulose membrane with a dry filter paper is overlaid on its upper surface.
- Then optimum pressure is exerted by placing some weight (0.5 kg) over the membrane causing a proper interaction between the two such that the dry filter draws the buffer through the gel.

Advanced Methods in Molecular Biology and Biotechnology. https://doi.org/10.1016/B978-0-12-824449-4.00020-7

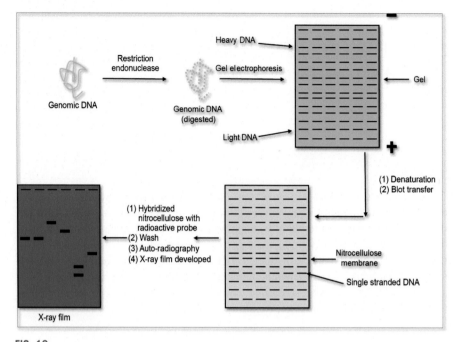

FIG. 16

Southern blotting.

- The nitrocellulose filter binds the DNA fragments that come into contact with it; it is subsequently baked at 80°C or given UV exposure leading to permanent fixing of DNA fragments to the Nitrocellulose filter.
- Then the filter is placed in a solution which contains a radiolabeled DNA probe. The probe hybridizes the complementary DNA on the Nitrocellulose filter.
- The hybridized DNA regions are obtained on an X-ray photographic film using autoradiography.

Applications

(i) Used in restriction fragment length polymorphism (RFLP) mapping.

(ii) Used in a phylogenetic study of different groups of organisms.

(iii) Used in identifying the gene rearrangements.

- Western blotting

    Western blotting, developed by **Towbin** et al., is a protein blotting technique which is used in the detection and analysis of protein of a particular specificity in a given sample. It works on the principle of antigen-antibody reaction (Fig. 17).

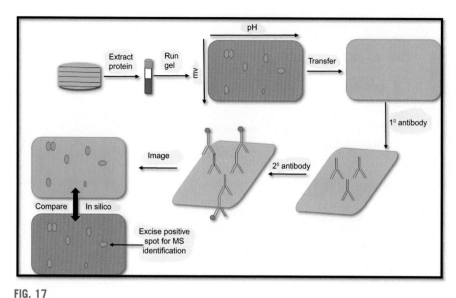

FIG. 17

Western blotting.

Step-by-step method details

It involves the following steps:

- Firstly, protein is extracted from a given sample.
- It is followed by the separation of protein by using sodium dodecyl sulfate (SDS) PAGE.
- Then the electrophoresed gel is transferred in a buffer and is given cold treatment for half an hour at 40°C.
- It is followed by nitrocellulose binding, i.e., binding of proteins onto the nitrocellulose filter.
- Then a soaked Whitman filter paper is placed on a cathode plate followed by a stack of coarse filter, Whitman filter, gel, nitrocellulose filter, Whitman filter, coarse filter, Whitman filter, and anode plate.
- The complete set is put in a transfer tank which is filled with sufficient transfer buffer.
- Then a current of 30 V overnight for 5 h is passed across causing protein migration from the gel to the nitrocellulose filter and getting bound on its surface.
- Then the use of radiolabeled antibodies of known structure hybridization of protein is carried out. These antibodies are responsible for identification of a specific amino acid sequence. The unhybridized antibodies are removed by washing the nitrocellulose filter in a wash solution.
- The hybridized sequences are detected autoradiographically.

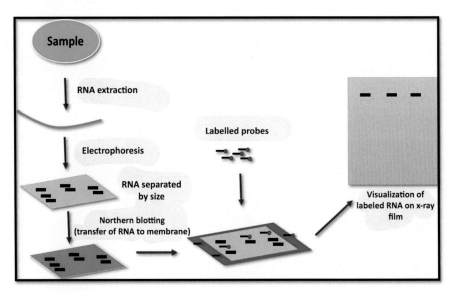

FIG. 18

Northern blotting.

Applications

(i) Used in clinical investigations.

(ii) Used to detect low-quantity proteins.

(iii) Used to ensure and quantify a gene product.

- Northern blotting

  Since the Southern blotting technique could not be used to blot RNA due to its inability to bind with the nitrocellulose filter, the Northern blotting technique was developed by Alwine et al. for detecting and analyzing RNA in a sample solution (Fig. 18).

Step-by-step method details

It involves the following steps:

- Firstly, mRNA from the target cells should be extracted and purified.
- Then separate these RNA segments via agarose gel electrophoresis containing formaldehyde that triggers RNA denaturation.
- Then this gel should be immersed in a depurination buffer for 6–10 min and washed gently with water.
- Then the gel containing RNA fragments should be transferred onto the aminobenzyloxymethyl cellulose paper (Reactive paper).

- The reactive paper bound with fragments is given gentle heat or exposed to UV rays to fix fragments properly.
- Now it is available for hybridization with a radiolabeled DNA probe and is subjected to it.
- After washing off the loosely bound probe, the mRNA-DNA hybrids are detected autoradiographically on an X-ray film.

Applications

(i) Used in screening
(ii) To study the gene expression
(iii) To investigate disease as a diagnostic technique.
   - Eastern blotting
     Eastern blotting, developed by **Bogdanov**, is used for the identification and analysis of carbohydrate epitopes, lipids, and Glycoconjugates. When blotted proteins are transferred onto the Nitrocellulose filter and are analyzed for posttranslational modifications using a specific probe, the carbohydrates and lipids could be analyzed.

Step-by-step method details

It involves the following steps:

- Firstly, using agarose gel electrophoresis, the targeted molecules are vertically separated.
- Then they are horizontally transferred onto the nitrocellulose filter.
- Then a primary antibody is added to the solution. They recognize the specific amino acid sequence on the basis of the specificity of the molecules.
- Then to remove loosely and unbound primary antibodies, it is washed.
- Then a labeled secondary antibody is added.
- The conformity of the molecule of interest is done by these labeled Probes.

Applications

(i) Used to detect protein modification.
(ii) Using ligands, it is used for binding studies.
(iii) Used in purifying phospholipids, glycolipids.

- North-Western blotting
  It is the hybrid analytical technique of northern blot and western blot used to detect the interactions between RNA and proteins.
- South-Western blotting
  It is the hybrid analytical technique of southern blot and western blot used to detect the interactions between DNA and proteins.

- Far-Eastern blotting
  This technique of blotting is used for the detection of lipid-linked oligosaccharides.
- Far-Western blotting
  This method of blotting is used for the detection of a particular protein immobilized on a blotting matrix.
- Dot blotting
  This technique of blotting is a special case of any of the above techniques where the analyte is directly added to the blotting matrix in the form of a "Dot" as opposed to separating the samples prior to blotting by electrophoresis.

Further reading

Ausubel, F. M., Brent, R., Kingston, R. E., et al. (Eds.). (1993). *Current protocols in molecular biology*. New York: John Wiley.

Dyson, N. J. (1991). Immobilization of nucleic acids and hybridization analysis. In T. A. Brown (Ed.), *Vol. II. Essential molecular biology: A practical approach* (pp. 111–156). UK: Oxford.

Sambrook, J., Fritsch, E. F., & Maniatis, T. (1989). In N. Ford, C. Nolan, & M. Ferguson (Eds.), *Molecular cloning: A laboratory manual* Cold Spring Harbor Protocols.

Southern, E. M. (1975). Detection of specific sequences among DNA fragments separated by gel electrophoresis. *Journal of Molecular Biology*, *98*(3), 503–517.

Towbin, H., Staehelin, T., & Gordon, J. (1979). Electrophoretic transfer of proteins from polyacrylamide gels to nitrocellulose sheets: Procedure and some applications. *Proceedings of the National Academy of Sciences*, *76*(9), 4350–4354.

Dot blot analysis

Definition and rationale

In a biological sample, a specific protein can be identified by the use of an immune detection system (e.g., a protein marker for a disease) and the technique is known as dot blot analysis or dot blotting. After the protein molecules become immobilized on a binding membrane (nitrocellulose or PVDF), these are probed with a primary antibody (specific for the protein of interest). Once bound, the antibody is analyzed, either with specific tags coupled to the primary antibody or with a secondary antibody. For example, if the primary antibody is a mouse antibody, the secondary antibody used will recognize all mouse antibodies. This step is followed by the step which helps to enhance the specificity of the dot blot technique. It prevents the occurrence of nonspecific interactions, i.e., sticking of antibodies to nonspecific proteins by placing a membrane in the mixture containing protein. Thus, the charges that would attract the antibodies are blocked by these protein molecules. Other blocking agents include milk powder, serum albumin, and casein. Nonanimal proteins are now being used in novel blocking agents which help to prevent any nonspecific interactions.

Before you begin

Sample spotting.

Materials and equipment

- Simulated Sample A (30 μL)
- Simulated Sample B (30 μL)
- IMU positive control (30 μL)
- IMU negative control (30 μL)
- MEM washing (1X)
- Blocking buffer (NAP-Blocker) (1X)
- BE Antibody-1
- BE Antibody-4 (Horseradish peroxidase secondary)
- HRP substrate (1 bottle)Protein-binding membrane (4 strips)
- Two tubes of 50 each
- Washing tray

Advanced Methods in Molecular Biology and Biotechnology. https://doi.org/10.1016/B978-0-12-824449-4.00021-9

Step-by-step method details

Timing: 4–6 h

Sample spotting
1. The first step involves the removal of a protective strip from the nitrocellulose membrane strip using forceps.
2. Take 5 μL of the sample and spot onto the membrane.
3. Similarly, spot all the samples on the membrane strip.
4. Then place all the membrane strips in a clean washing tray.

Detection of protein
5. Add 1X Blocking Buffer @ 20 mL to block the nonspecific sites.
6. Incubate the membrane for 30–60 min at room temperature while shaking gently.
7. Then prepare a primary antibody by adding 40 μL BE Antibody-1 to 1X Blocking Buffer (20 mL).
8. Pour off the Blocking Buffer.
9. Add the primary antibody solution to the membrane and incubate again.
10. Pour off the antibody solution and wash 3 times with 1X MEM Washing Buffer @ 20 mL for 10 min.
11. Mix 40 μL BE Antibody-4 (HRP Secondary) and 1X Blocking Buffer (20 mL) in a 50 mL tube for the preparation of the secondary antibody.
12. Add the secondary antibody solution to the membrane and incubate while shaking gently.
13. Pour off the antibody solution and wash 3 times with 1X MEM Washing Buffer @ 20 mL for 10 min.
14. Now, add 5 mL of HRP substrate to the membrane.
15. Shake the membrane at room temperature for 5 min or until color develops.
16. Discard the substrate and add distilled ionized water to stop the color reaction.

Expected outcomes

Darker dots indicate more protein (Fig. 19).

Advantages

This technique does not demand the use of electrophoresis or separation of bands on solid media. The presence of genes in a sample can be detected from a single test run. The transfer of biomolecules from a gel to a filter membrane is also not required. Dot blot technique involves the direct blotting of a biomolecule onto the membrane.

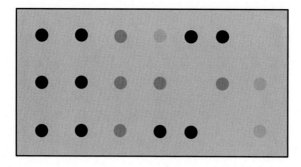

FIG. 19

Dot blot analysis.

Limitations

This technique has a limitation as it does not provide any information about the size and molecular weight of the identified biomolecules.

Further reading

Heinicke, E., Kumar, U., & Munoz, D. G. (1992). Quantitative dot blot assay for proteins using enhanced chemiluminescence. *Journal of Immunological Methods, 152,* 227–236.

Jahn, R., Schiebler, W., & Greengard, F. (1984). A quantitative dot-immunobinding assay for proteins using nitrocellulose membrane filters. *Proceedings of the National Academy of Sciences of the United States of America, 81,* 1684–1687.

Stott, D. I. (1989). Immunoblotting and dot blotting. *Journal of Immunological Methods, 119,* 153–187.

ELISA (enzyme-linked immunosorbent assay)

22

Definition

An immune-chemical technique used for the confirmation of the presence of a specific protein, i.e., antigen or antibody in a given particular biological sample and its quantification is known as ELISA. It is a more specific and sensitive assay than other immunoassays for the detection of a biological molecule. It is also known as a *solid-phase enzyme immunoassay* as the presence of a specific protein is ascertained by using an enzyme-linked antigen or antibody as a marker. A colored reaction product is generated when an enzyme conjugated with an antibody reacts with a colorless substrate. The various enzymes that have been employed for ELISA include horseradish peroxidase, alkaline phosphatase, β-galactosidase, etc.

Rationale

The ELISA technique involves the following three principles in combination:

- Antigen-antibody reaction (**Immune reaction**).
- Formation of chromogenic product from colorless substrate (**Enzymatic chemical reaction**).
- Detection and measurement of color intensity of the colored products (**Signal detection and Quantification**).

Types of ELISA

- **Direct ELISA**

  A target protein (or a target antibody) is immobilized on the surface of microplate wells and incubated with an enzyme-labeled antibody to the target protein (or a specific antigen to the target antibody). After washing, the activity of the microplate well-bound enzyme is measured. A schematic representation of Direct ELISA is given in Fig. 20A.

- **Indirect ELISA**

  This type of ELISA is used for the detection of an **antibody** in a biological sample. The antigen is adhered to the wells of the microtiter plate. The primary antibody binds specifically to the antigen after the addition of the sample. The solution is then

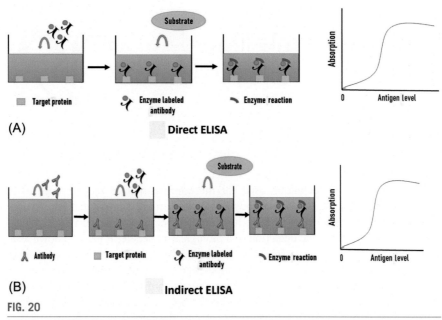

FIG. 20

(A) Direct ELISA. (B) Indirect ELISA.

washed to remove the unbound antibodies, the enzyme is conjugated, and secondary antibodies are added. The primary antibody is then quantified through a color change after the substrate for the enzyme is added. The concentration of primary antibody present in the sample is directly correlated with the intensity of the color. Schematic representation of Direct ELISA is given in Fig. 20B.

Materials and equipment

- Pipettes.
- Washing system.
- ELISA plate reader.
- 96-Well plates coated with either inactivated antigen or antibody.
- Sample diluents.
- Controls. (Negative and positive controls which help to normalize or standardize the plates.)
- Buffered wash concentrate.
- Enzyme-labeled antibodies (Conjugates).
- Substrate.
- Stop solution. (It stops the enzyme substrate reaction and color development.)

Step-by-step method details
Timing: 5–6 h

- Prepare the reagents and equipments needed.
- Add the diluted antibody to each well of a 96-well ELISA plate.
- Seal the plate to prevent evaporation, and incubate it at 4°C for 15–18 h to immobilize the antibody.
- Wash the diluted antibody 3 times with washing solution.
- Add blocking buffer to each well, and incubate it at 37°C for 1 h to reduce nonspecific binding of the target protein to the well.
- Wash the blocking buffer 3 times with washing solution.
- Add sample dilution buffer to dilute the samples, and then add 100 µL of each sample to each well.
- For the calibration curve, prepare a dilution series of the standard on the same plate and incubate it at 37°C for 1 h.
- Then wash the samples with washing solution 5 times.
- Add sample dilution buffer to dilute the detection antibody and add 100 µL to each well and incubate at 37°C for 1 h.
- Remove the detection antibody after reaction, and wash 5 times with washing solution.
- Add sample dilution buffer to dilute an enzyme-labeled secondary antibody, and add 100 µL to each well and incubate at 37°C for 1 h.
- Remove the secondary antibody after reaction, and wash 5 times with washing solution.
- Add a substrate solution and incubate as the color develops.
- When the color has sufficiently developed, add a stop solution for stopping the reaction.
- Measure the absorption at 450 nm using a plate reader.

Quantification and statistical analysis

The ELISA produces three different types of data output:

- **Quantitative:** Here ELISA data can be interpreted in comparison to a standard curve (a serial dilution of a known purified antigen) so as to calculate the concentrations of antigen in various samples precisely.
- **Qualitative:** Here ELISAs can be used to obtain a yes or no answer which indicates whether a particular antigen is present in a sample, comparing it to a blank well containing no antigen or an unrelated control antigen.
- **Semiquantitative:** Here ELISAs can be used to compare the relative levels of antigen in assay samples, since the intensity of the signal varies directly with antigen concentration.

Advantages
Direct ELISA

- Quick method because it involves fewer steps and uses only one antibody.
- Elimination of cross-reactivity of secondary antibody.

Indirect ELISA

- A number of labeled secondary antibodies are commercially available.
- A number of primary antibodies can be made in one species and the same labeled secondary antibody can be used for detection; hence, it is versatile.
- As a primary antibody, its maximum immune-reactivity is retained.
- The sensitivity of this technique is increased because of the presence of several epitopes in each primary antibody that can be bound by the labeled secondary antibody, allowing for signal amplification.

Sandwich ELISA

- High specificity because the antigen/analyte is specifically captured and detected.
- The method is also suitable for crude/impure samples because the antigen does not require purification prior to measurement.
- The method is both flexible and sensitive (direct as well as indirect detection methods can be used).

Competitive ELISA
- Highly sensitive method that can be used even when relatively small amounts of the specific detecting antibody are present.

Limitations
Direct ELISA

- Labeling with reporter enzymes or tags might adversely affect immunoreactivity of the primary antibody
- It is a time-consuming and expensive method as it requires labeling of primary antibodies for each specific ELISA system
- A limited number of conjugated primary antibodies are commercially available.
- No flexibility in choice of primary antibody label from one experiment to another.
- Signal amplification is minimal.

Indirect ELISA
Occurrence of nonspecific signal due to cross-reactivity with the secondary antibody.
Sandwich ELISA
Requires more optimization to identify antibody pairs and to ensure there is limited cross-reactivity between the capture and detection antibodies.

Further reading

Alhajj, M., & Farhana, A. (2020). Enzyme linked immunosorbent assay (ELISA). In *Stat Pearls [Internet]* StatPearls Publishing.

Aydin, S. (2015). A short history, principles, and types of ELISA, and our laboratory experience with peptide/protein analyses using ELISA. *Peptides, 72*, 4–15.

Kohl, T. O., & Ascoli, C. A. (2017a). Direct competitive enzyme-linked immunosorbent assay (ELISA). *Cold Spring Harbor Protocols, 2017*(7). pdb-prot093740.

Kohl, T. O., & Ascoli, C. A. (2017b). Indirect immunometric ELISA. *Cold Spring Harbor Protocols, 2017*(5). pdb-prot093708.

Kohl, T. O., & Ascoli, C. A. (2017c). Immunometric double-antibody sandwich enzyme-linked immunosorbent assay. *Cold Spring Harbor Protocols, 2017*(6). pdb-prot093724.

Konstantinou, G. N. (2017). Enzyme-linked immunosorbent assay (ELISA). In *Food Allergens* (pp. 79–94). New York, NY: Humana Press.

Nakane, P. K., & Pierce, G. B., Jr. (1966). Enzyme-labeled antibodies: preparation and application for the localization of antigens. *Journal of Histochemistry and Cytochemistry, 14*(12), 929–931.

Study of principle of centrifugation

23

Definition

Centrifugation is a technique that helps to separate chemical mixtures by the application of centrifugal force, and the device that is used to separate out the particles [macromolecules (proteins and nucleic acids), micromolecules, cells, subcellular organelles, viruses, etc.] from a mixture of particles depending on their size, shape, viscosity, density, and rotor speed is called a ***Centrifuge***.

The centrifugal force produced by the centrifuge is measured by the gravity units as

$$\omega = \frac{2\pi * rev/min}{60}$$

where

G = centrifugal field
ω = angular velocity of the rotor
r = radial distance of the particle from the center of the rotation

Again, angular velocity of the rotor is given by the formula;

- It works on the basic principle of sedimentation.
- The gravitational acceleration (g) is replaced by the centrifugal force in the presence of a centrifugal field. Under the influence of gravitational force (g-force), particles separate as per their size, shape, viscosity, density, and centrifugal force. The simple example is of a spherical molecule. If the liquid has the density of—ρ_L and the particle has a density of ρ_M and if $\rho_M > \rho_L$, then the molecule will sediment.
- In a laboratory centrifuge, the centripetal acceleration causes the heavier substances to settle down to the bottom of the container or to move outward in the radial direction, whereas the low-density substances either rise to the top or are displaced to the center (Fig. 21).

Advanced Methods in Molecular Biology and Biotechnology. https://doi.org/10.1016/B978-0-12-824449-4.00023-2

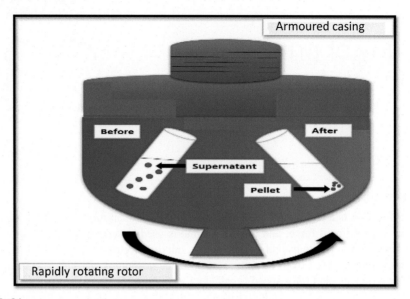

FIG. 21

Centrifuge.

Types of centrifuge

- **Low-speed centrifuge**
  - The speed of a low-speed centrifuge ranges from 4000 to 5000 rpm.
  - It is the centrifuge commonly used in the laboratories for routine sedimentation of heavy particles.
  - This instrument usually operates at room temperature without any temperature control unit.
  - This type of centrifuge is used for the sedimentation of RBCs as long as the particles get tightly packed into a pellet and the supernatant is then separated by decantation.
  - The following types of rotors are used;
    - Fixed angle
    - Swinging bucket.
- **High-speed centrifuge**
  - The speed of a high-speed centrifuge ranges from 15,000 to 20,000 rpm.
  - It is commonly used in more sophisticated biochemical applications.
  - In this instrument, speed and temperature can be controlled manually as required for biological samples.
  - The following types of rotors are used in it:
    - Fixed angle
    - Swinging bucket
    - Vertical rotors

- **Ultracentrifuge**
  - An ultracentrifuge has a high speed of 65,000 rpm.
  - It is the most sophisticated instrument.
  - Due to high speed, intense heat is generated in this type of centrifuge; therefore, it is important to refrigerate the spinning chambers and keep the chambers at high vacuum.
  - It is used for preparative work as well as analytical work.

Before you begin

Preparation of different samples for centrifugation

Materials and equipment

- Centrifuge
- 2-mL microfuge tubes
- Sterile water
- Micropipettes
- Samples

Step-by-step method details

- **Differential centrifugation (differential pelleting):**
  - It is the most simple and commonly adopted type of centrifugation.
  - In this, the tissue is homogenized in a buffer containing sucrose solution at 32°C.
  - Then, at a constant temperature, the homogenate obtained is allowed to spun.
  - After a couple of minutes, a pellet is formed at the bottom of the centrifuge tube.
  - The supernatant is transferred to a new centrifuge tube which is then placed on a rotating shaker at high speed in the following steps.
- **Density gradient centrifugation:**
  - It is mainly used for the purification of viruses, ribosomes, membranes, etc.
  - In this type of centrifugation, a density gradient (e.g., sucrose) is created by overlaying lower concentrations of solute on higher concentrations in tubes.
  - The particles of choice are then placed on top of the solute and spun in ultracentrifuges.
  - The particles of choice start traveling through the gradient unless and until they reach a point where the density of particles equals the density of the solute (sucrose).
  - The fraction is separated and finally analyzed.
- **Band/rate zonal density gradient centrifugation:**
  - It is based on the notion of movement of sediments through the liquid medium (*Sedimentation coefficient*).

- In this type, a density gradient is created in a centrifuge tube with a particular solute (e.g., sucrose).
- The sample of choice (e.g., protein) is kept on the top of the gradient and then centrifuged.
- During centrifugation, faster-sedimenting particles in the sample move ahead of slower ones, i.e., samples separate into different zones in the solution gradient.
- The proteins settle down (i.e., sediment) on the basis of their sedimentation coefficients and the fractions are collected in a tube by creating a hole at the bottom.
- **Isopycnic centrifugation:**
  - In this technique, in a centrifuge tube a biological sample is loaded with a gradient forming solution (cesium salt).
  - The sample and cesium salt solution is uniformly distributed in a tube and centrifuged in an ultracentrifuge.
  - During centrifugation, the cesium salts redistribute to develop a density gradient from top to bottom.

Expected outcomes

The molecules with a lower molecular weight settle down at the bottom, whereas the molecules with a higher molecular weight float at the top.

Advantages

The technique causes separation of two completely solubilized substances. It separates chalk powder from water. It causes the separation from an airflow by using a method of cyclonic separation. The technique helps in analyzing the hydrodynamic properties of various large molecules. It helps in purification of animal cells. It causes fractionation of subcellular organelles, membranes/membrane fractions. It helps not only in production of skimmed milk by the removal of fat from milk but also in stabilization and clarification of wine. It helps in separating urine and blood components in various research laboratories. It also helps in the separation of proteins using the salting out technique of purification. The technique helps in DNA and protein extraction from plants, animals, etc.

Limitations

The technique has high initial costs. The device used in centrifugation requires high energy consumption and it is a batch process.

Optimization and troubleshooting
Problem
There is wobbling or shaking of the centrifuge and improper sedimentation is also observed.

Potential solution
During the shaking of the centrifuge, it should be unplugged. Balance should be maintained for the samples inside the centrifuge to avoid any kind of improper sedimentation.

Safety considerations and standards
- Always mix the samples gently.
- Do not open the upper lid of the centrifuge until it stops completely.

Alternative methods/procedures
Cesium chloride density gradient centrifugation.
In this method of centrifugation, the particles are separated on the basis of their density.

Further reading
Fisher, D., Francis, G. E., & Rickwood, D. (1998). *Cell separation: A practical approach.* Oxford: Oxford University Press.

Graham, J. M., & Rickwood, D. (1997). *Subcellular fractionation: A practical approach.* Oxford: IRL Press at OUP.

Rickwood, D. (1984). *Centrifugation: A practical approach* (2nd ed.). Oxford: IRL Press at OUP.

Rickwood, D. (1992). *Preparative centrifugation: A practical approach.* Oxford: IRL Press at OUP.

Study of principle of chromatography

24

Definition

Most of the chemical interactions produce products in a mixture state containing both useful and unwanted molecules. Chromatography is an investigative or analytical technique that allows workers not only to separate but also to identify the chemical constituents present in a mixture so that the desired constituents could be easily separated and thoroughly studied. Tswett in 1903–06 invented the technique of chromatography, developed chromatogram using pure solvent, devised nomenclature, and also used mild adsorbents to resolve photosynthetically active chloroplast pigments.

Chromatography works on the principle of differential partitioning between the mobile and stationery phases. Basically, different components of the analyte mixture have differential affinities (strength of adhesion) toward the mobile and stationary phases resulting in the differential separation of the components. However, the accord of individual components depends on two properties, i.e., adsorption and solubility. Adsorption is the property of compounds to adhere to the surface of the stationery phase molecules without forming a solution. Solubility in turn is the property of compounds to get dissolved in the mobile phase and form a solution. It has been observed that:

- The components with a higher degree of adsorption toward the stationary phase have a slower pace through the column.
- The components with a higher degree of solubility in the mobile phase move rapidly between the column.

Different types of chromatographic techniques

- TLC—thin layer chromatography
- Gel filtration/size exclusion chromatography
- Ion exchange chromatography
- Affinity chromatography
- Paper chromatography

Advanced Methods in Molecular Biology and Biotechnology. https://doi.org/10.1016/B978-0-12-824449-4.00024-4

Terms	Definition
Mobile phase/carrier	It refers to the solvent molecules moving through the column
Stationary phase/adsorbent	It refers to the substances that stays inside the column and fixed
Eluent	Refers to the fluid which enters the column
Eluate	Refers to the fluid which exits the column (that is collected in flasks)
Elution	It is the process of washing out a compound through a column with the help of a suitable solvent
Analyte	Refers to the mixture whose individual components have to be separated and analyzed

Expected outcomes

Different compounds separate as per their differential affinities and get collected in separate beakers (Fig. 22).

Advantages

The technique is cheap and simple with high reliability.

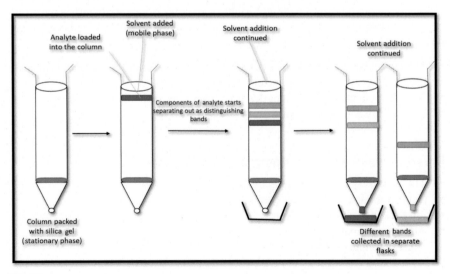

FIG. 22

Process of chromatography.

Limitations

It has a slightly higher separation time.

Further reading

Biosciences, A. (2002). *Ion exchange chromatography, principles and methods* (p. 751). Amercham Pharmacia Biotech SE.

Cuatrecasas, P., Wilchek, M., & Anfinsen, C. B. (1968). Selective enzyme purification by affinity chromatography. *Proceedings of the National Academy of Sciences of the United States of America, 61*(2), 636.

Das, M., & Dasgupta, D. (1998). Pseudo-affinity column chromatography based rapid purification procedure for T7 RNA polymerase. *Preparative Biochemistry & Biotechnology, 28*(4), 339–348.

Wilchek, M., & Chaiken, I. (2000). An overview of affinity chromatography. In *Affinity chromatography* (pp. 1–6). Humana Press.

Thin layer chromatography (TLC)

Definition

TLC technique is based on another chromatographic technique known as *Planar chromatography*. This technique is mostly adopted by researchers in various fields of chemistry, for the identification of substances in the mixture of compounds, e.g., amino acids. It is a semiquantitative method of analysis. *High-performance thin layer chromatography* (HPTLC) is the most sophisticated and precise quantitative version of TLC.

Rationale

The chromatographic method is again based on the principle of separation, i.e.,

- Components are separated based on the relative affinity of compounds toward the adsorbent and the carrier phase.
- Compounds undergo capillary action, i.e., compounds travel over the surface of the adsorbent phase under the influence of the carrier phase.
- Compounds with a higher affinity to the adsorbent phase travel slowly while the others travel faster resulting in the separation of components in the mixture of compounds.
- Once the components separate, the individual components can then be visualized as spots at the respective level of distance traveled by them on the plate.

Materials and equipment

- **TLC plates**—The surface of these plates is coated with a thin layer of the stationary phase. These plates are heat stable and chemically inert.
- **TLC chamber**—The chamber which maintains a uniform environment inside the TLC plate for the proper development of spots and also avoids the evaporation of solvents, while keeping the whole process dust free.
- **Mobile phase**—It consists of a solvent or a mixture of solvents. The mobile phase used should be particulate-free, chemically inert with the stationary phase, and of the highest purity for proper development of TLC spots.

Advanced Methods in Molecular Biology and Biotechnology. https://doi.org/10.1016/B978-0-12-824449-4.00025-6

- **Filter paper**—It helps in the development of a uniform rise in a mobile phase over the length of the stationary phase.

Step-by-step method details

- Spread the adsorbent phase uniformly onto the TLC plate and allow it to dry and stabilize.
- Make a mark at the bottom of the TLC plate with a pencil for applying the sample spots.
- Apply the sample solutions on the spots.
- Pour the carrier phase to a level few centimeters above the bottom of the TLC chamber.
- Cover the inner walls of the TLC chamber with moistened filter paper to maintain equal humidity.
- Immerse the prepared plate with sample spotting in the mixture of solvents in a TLC chamber for a sufficient period of time for the development of spots and then close the chamber with a lid.
- After a couple of minutes, remove the plates and allow them to dry.

The sample spots can then be visualized in a suitable UV light chamber, or in an Iodine jar or by any other method as recommended.

Quantification and statistical analysis

Retention factor (R_f):

- Defined as the ratio of the distance traveled by the individual component to the total distance traveled by the solvent (Fig. 23)
- The components having a lower R_f value are considered to be more polar (i.e., less soluble), and vice versa.

Distance traveled by the components on the TLC plate

Name of the component	Distance traveled by the component (cm)	Distance traveled by the solvent (cm)	Retention factor of the component (R_f)
A	3.5	6	$R_f a = 3.5/6 = 1.7$
B	3	6	$R_f b = 3/6 = 0.5$
C	2	6	$R_f c = 2/6 = 0.33$
D	1	6	$R_f d = 1/6 = 0.16$

On the basis of the R_f value as calculated above, component d is considered to be the most polar and component a the least polar one.

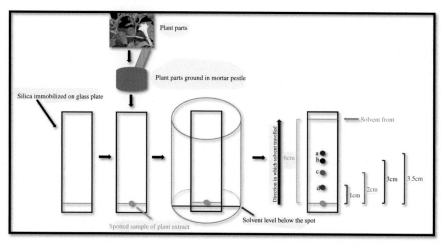

FIG. 23

Thin layer chromatography.

Advantages

It is a simple, cheap, and less time-consuming process. It helps in easy visualization of separated compound spots and identification of the individual components such as acids, alcohols, alkaloids, amines, antibiotics, proteins, etc. TLC allows the isolation of most of the compound mixtures. It helps in easy assessment of the purity standards of a given sample.

Limitations

TLC plates lack a longer stationary phase. The length of the separation is limited compared to other chromatographic techniques. The results obtained from TLC are difficult to reproduce. TLC operates as an open system; therefore, some factors such as humidity and temperature can affect the final outcome of the chromatogram. TLC cannot be used for lower detection limits. It is only a qualitative analysis technique and not a quantitative one.

Safety considerations and standards

Do not keep acetone near open flames and do not breathe the fumes. Be careful while using dyes in order to avoid staining clothing and hands.

Alternative methods/procedures

- Gel Filtration or Size-Exclusion Chromatography
- Ion Exchange Chromatography.
- Affinity Chromatography.
- Paper chromatography

Applications

- In checking the purity standards of given samples.
- In identification of various macromolecules such as acids, alcohols, amines, alkaloids, antibiotics, proteins, etc.
- In the evaluation of the reaction process, by the assessment of intermediates, reaction course, and so forth.
- In the purification of various samples.
- In keeping a check on the performance of other kinds of separation processes.

Further reading

Kim, Y. J. (1998). A simple streaking device for preparative thin layer chromatography. *Journal of Chemical Education, 75*(5), 640.

Sherma, J., & Fried, B. (Eds.). (2003). *Vol. 89. Handbook of thin-layer chromatography* CRC Press.

Waksmundzka-Hajnos, M., Petruczynik, A., Hajnos, M.Ł., Tuzimski, T., Hawryl, A., & Bogucka-Kocka, A. (2006). Two-dimensional thin-layer chromatography of selected coumarins. *Journal of Chromatographic Science, 44*(8), 510–517.

Gel-filtration or size-exclusion chromatography 26

Definition

Gel-filtration or size-exclusion chromatographic technique is primarily used for analytical assays and semi-preparative purifications. The variations in the size of the molecules help in this technique and they are separated solely on this basis. **Grant Henry Lathe** and **Colin R Ruthven** were the first who used this method for the separation of various analytes of different sizes with starch gels as the matrix. Later, **Jerker Porath** and **Per Flodin** recommended the use of dextran gels as matrices for the separation of analytes. Other gel-filtration matrices include agarose and polyacrylamide.

Rationale

It is the method of separating various mixtures according to the molecular size or hydrodynamic volume of the analytes. As the solutes pass through the stationary phase consisting of the pores of various sizes and cross-linked polymeric gels or beads, the separation is attained by the differential exclusion or inclusion of solutes.

The whole process of separation is based on different permeation rates of every solute molecule into interior of the gel particles. The particles of smaller size move through the pores and travel through a longer path and get eluted last, whereas the particles of larger size move only through few pores and hence travel through a shorter path and get eluted first. The particles are solely separated according to their sizes, therefore the name *Size-Exclusion Chromatography.*

- It is also referred to as *Gel-Filtration Chromatography* **(GFC)** when used for the separation of biomolecules in aqueous systems, and *Gel Permeation Chromatography* **(GPC)** when used for the separation of organic polymers in nonaqueous systems.

Materials and equipment

- Columns
- Gel beads
- Protein solution

Advanced Methods in Molecular Biology and Biotechnology. https://doi.org/10.1016/B978-0-12-824449-4.00026-8

Step-by-step method details

Timing: 1–2 h

- The gel powder is immersed in the solvent and allowed to swell (mobile phase).
- The solvent gets absorbed by the powder and bulges to form a gel matrix having molecular sieves to allow smaller particles to pass.
- The gel is then packed into the column to an appropriate height.
- The mixture which is to be separated is mixed in the mobile phase.
- The mixture obtained is then poured over the top of the column.
- The mixture is finally permitted to percolate through the gel matrix in the column.
- The different eluates are then collected in different beakers at their appropriate times.

Expected outcomes

The particles of smaller size move through all the pores and travel a longer path and get eluted last, whereas the particles of larger size, move only through few pores and hence travel shorter path and get eluted first (Fig. 24)

Advantages

The technique helps in determining the molecular weight distribution of various types of polymers. It helps in the purification of samples and in the separation of organic compounds like polypeptides, polystyrenes, etc.

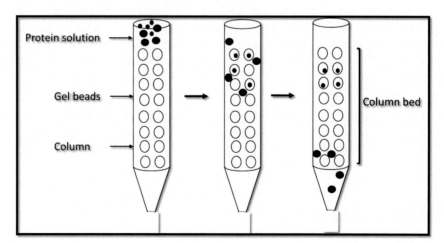

FIG. 24

Size-exclusion chromatography.

Alternative methods/procedures

- Thin-layer chromatography
- Ion-exchange chromatography
- Affinity chromatography
- Paper chromatography

Further reading

Determann, H. (2012). *Gel chromatography gel filtration gel permeation molecular sieves: A laboratory handbook*. Springer Science & Business Media.

Kostanski, L. K., Keller, D. M., & Hamielec, A. E. (2004). Size-exclusion chromatography—a review of calibration methodologies. *Journal of Biochemical and Biophysical Methods*, *58*(2), 159–186.

Kuk, C., Kulasingam, V., Gunawardana, C. G., Smith, C. R., Batruch, I., & Diamandi Berek, D. (2004). Critical assessment of the size exclusion chromatography. *Polimery*, *5*, 311–386.

Mojsiewicz-Pieńkowska, K. (2014). Size exclusion chromatography a useful technique for speciation analysis of polydimethylsiloxanes. In *Green Chromatographic Techniques* (pp. 181–202). Dordrecht: Springer.

Ion-exchange chromatography

27

Definition

Ion-exchange chromatography (IEC) is used for the separation of ionizable molecules according to differences in charge properties. This is the more commonly used separation technique than affinity chromatography.

Rationale

The charged molecules in the sample are separated with the help of electrostatic forces of attraction which are passed through the ionic resin at particular pH and temperature. The separation takes place by *reversible exchange of ions* which are present in the solution and in the ion exchange resin. The process of separation of molecules from the mixture is highly dependent on the type of ion exchange resin used.

Ionic resins are of two types as:

- *Cation-exchange resins*—They are negatively charged exchangers and contain positively charged counter ions and are meant for retaining cations in the column, hence named so. They could be strongly acidic, intermediate acidic, or weakly acidic in nature.
- *Anion-exchange resin*—They are positively charged exchangers and contain negatively charged counter ions and are meant for retaining anions in the column, hence named so. They could be either strongly basic, intermediate basic, or weakly basic in nature (Fig. 25).

Ion-exchange chromatography is a powerful method used in separating two protein species which resemble one another closely but have a difference in at least one charged amino acid. In this type of chromatography, the ionic molecules get stuck in the matrix resin and are retained on an ion exchanger because of their reversible interaction with the oppositely charged group on the ion exchanger.

Before you begin

Preparation of the sample and the column.

Advanced Methods in Molecular Biology and Biotechnology. https://doi.org/10.1016/B978-0-12-824449-4.00027-X

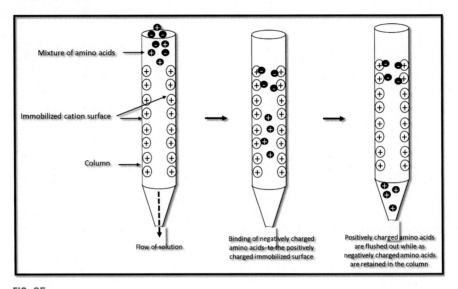

FIG. 25

Anion-exchange chromatography.

Materials and equipment

- Protein sample
- 0.3 mL column
- Equilibration buffer (pH 8.1) centrifuge
- Centrifuge
- Sterile water
- Micropipettes
- Test tubes

Step-by-step method details
Preparing the sample and the column

- Firstly, add 0.2 mL of equilibration buffer (pH 8.1) to the protein sample that is under consideration for separation and mix it thoroughly.
- To remove any froth present in the sample, centrifuge it for 2 min.
- After making sample froth free, place the cation-exchange column in a test tube for 5 min in order to allow the resin present in it to settle.
- Then ensure the test tube is upright by placing it onto a ring stand with the help of a column.

- Permit the buffer in the column to drip out at its own into the test tube under gravity by opening first the top cap of the column and then the bottom cap carefully.
- Wash the column twice with a column-volume of equilibration buffer (here the column volume is 0.3 mL).
- Suppose before adding the second column-volume, the first column-volume (0.3 mL) of equilibration buffer oozes out drop by drop into the waste vial. This step allows the column to equilibrate with the equilibration buffer. Without disturbing the resin, add the equilibration buffer slowly.

Running a protein sample through a cation-exchange column

- Label a centrifuge tube of 2 mL as "Unbound 1" and put the column in it. Carefully load 0.1 mL of the protein sample onto the top of the column placed in the said tube.
- Once the sample has been loaded, it should be washed with 0.3 mL of equilibration buffer by adding the buffer at the top of the column and permitting it to drip all the way through.
- The above step should be repeated 4 times and the eluent should be collected each time in separate centrifuge tubes of the same dimensions, Label the tubes serially as "Unbound 1 to 5."
- For the last two washes, all unbound species should come off the column to ensure that it is centrifuged for 10 s at $1000 \times g$.
- Now put the column in a new collection tube of 2 mL and label it "Bound 1." On top of the column, carefully add 0.3 mL of elution buffer (pH 5.5).
- It should be followed with centrifugation of the resulting eluent for 10 s at $1000 \times g$.
- By placing the column in a new centrifuge tube repeat the elution step 2 times again by adding 0.3 mL of it, and centrifuging for 10 s. Label these tubes as "Bound 2 and 3."

Keenly observe and record any color changes if occurs. Hemoglobin (Hb) looks reddish-brown in color while cytochrome C pigment simply looks reddish.

Expected outcomes

Cation-exchange (ionic) of hemoglobin and cytochrome C pigments

- Hemoglobin has a pH of 6.8 and cytochrome C has a pH of 10.5. Thus when an equilibration buffer of pH 8.1 is used, cytochrome C binds to the column because of its positive charge whereas hemoglobin does not bind to the column because of its negative charge.
- Since hemoglobin is not bound to the column hence visualized in the unbound fractions.
- As cytochrome C is bound to the cation exchange column hence visualized in bound fractions.

Advantages

It is used for deionization and softening of hard water in various commercial and research activities. It is used to make solutions ion free hence used in their purification. In biochemistry, it is used for the separation of metabolites from urine, blood, drugs, etc., it is also used in the separation of organic compounds like proteins, carbohydrates, nucleotides present in the mixture. The technique is also used in purifying the enzymes obtained from the tissue of the living organisms.

Alternative methods/procedures

- Gel filtration or size-exclusion chromatography
- Thin-layer chromatography
- Affinity chromatography
- Paper chromatography

Further reading

Fritz, J. S. (2004). Early milestones in the development of ion-exchange chromatography: A personal account. *Journal of Chromatography A, 1039*(1–2), 3–12.

Katoch, R. (2011). *Analytical techniques in biochemistry and molecular biology.* Springer Science & Business Media.

Kent, U. M. (1999). Purification of antibodies using ammonium sulfate fractionation or gel filtration. In *Immunocytochemical methods and protocols* (pp. 11–18). Humana Press.

To perform affinity chromatography

28

Definition and rationale

This technique is similar to liquid chromatography and is employed for the separation and analysis of various sample components (analytes) by the use of a reversible biological interaction and/or molecular recognition, e.g., enzyme with an inhibitor and antigen with an antibody.

It is based on the following principle: The adsorbent phase consists of a supporting matrix (e.g., cellulose beads) upon which the substrate bounds covalently, in such a way that the reactive groups important for enzyme binding get exposed. As soon as the crude mixture of proteins passes through the chromatography column, proteins with binding sites for the immobilized substrate binds to the adsorbent phase, and all other proteins are eluted from the column through the void volume. Once all the proteins are eluted out, the bound enzyme(s) can be then eluted out by employing various methods.

Key resources table

Reagent or resource	Source	Identifier
• Albumin affinity column	Sigma-Aldrich	GE28-9466-03
• Animal serum albumin (ASA) vial of 100 μL	Sigma-Aldrich	MFCD00162221
• Affinity binding buffer (2 mL)	Sigma-Aldrich	SAB2501882
• Affinity elution buffer (1 mL)	Sigma-Aldrich	I5513
RED660 reagent	Sigma-Aldrich	–

Materials and equipment

- Albumin affinity column
- Animal serum albumin (ASA) vial of 100 μL

- Affinity binding buffer (2 mL)
- Affinity elution buffer (1 mL)
- RED660 reagent
- Centrifuge tubes (2 mL)
- Centrifuge

Step-by-step method details

- Firstly, centrifuge the affinity column for 5 s at 500 × g to bring the resin to the bottom of the column.
- Place the affinity column in a 2-mL centrifuge tube and centrifuge for 5 s to let the buffer drain out after breaking off the bottom plug of the column.
- Then, place the column in another 2 mL centrifuge tube, followed by the addition of 0.2 mL affinity binding buffer to the column.
- Now, centrifuge it for 5 s to let the buffer drain into the 2 mL centrifuge tube.
- Discard the buffer obtained in the centrifuge tube.
- Repeat steps 3, 4, and 5 twice to thoroughly equilibrate the column.
- Place the column in a new 2 mL centrifuge tube.
- Load 100 μL of ASA (animal serum albumin) to the column.
- Incubate the sample for 5 min, followed by centrifugation for 5 s.
- Collect the sample in another centrifuge tube and label this as *Fraction 1*.
- With the help of affinity binding buffer, wash the column 3 times by placing the column in a clean centrifuge tube and add affinity binding buffer of 2 mL to the column, followed by centrifugation for 5 s and label this tube as *Fraction 2*.
- Repeat step 11 (2 times) in two separate tubes and label them as *Fraction 3* and *Fraction 4*, respectively.
- The sample is then eluted with the help of a salt gradient.
- Place the column into a new centrifuge tube for gradient elution.
- Now add 0.2 mL of the lowest salt concentration buffer and label this as *Fraction 5* and centrifuge it for 5 s.
- Collect the above fraction in another 2 mL collection tube.
- Place the column in a fresh and clean centrifuge tube.
- Starting from *Fraction 6* to *Fraction 10*, add the next elution buffer and repeat step 8 technique until all the elution buffers have been added.
- All the six fractions are then collected in six separate 2 mL centrifuge tubes.

Advantages

It is used in the purification of nucleic acids and proteins from cell-free extracts and blood.

Limitations

The ligands used for affinity chromatography are very costly, also sometimes leakage can be observed. Low-productivity and nonspecific adsorption can inhibit the use of affinity chromatography in clinical research.

Safety considerations and standards

- It is important to store the affinity column and proteins at 4 °C while all other reagents could be stored at room temperature.
- Centrifuge the small vials before opening to prevent wastage of reagents.
- Use excess of ligand over the target protein in a minimal resin volume to maximize yield and minimize nonspecific binding.
- To minimize protein losses due to absorption, use polypropylene or polyethylene tubes instead of polystyrene tubes.
- Contaminants must be removed before protein purification.
- Ensure that the affinity ligand binds the target actively after the immobilization procedure (Fig. 26).

Alternative methods/procedures

- Gel filtration or sSize-exclusion chromatography
- Thin-layer chromatography

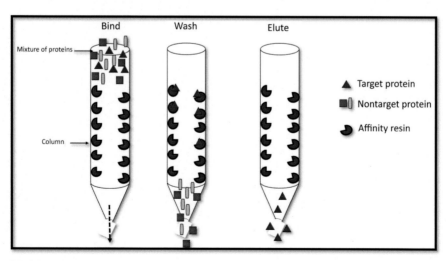

FIG. 26

Affinity chromatography.

- Ion-exchange chromatography
- Paper chromatography

Further reading

Hage, D. S., Anguizola, J. A., Li, R., Matsuda, R., Papastavros, E., Pfaunmiller, E., et al. (2017). Affinity chromatography. In *Liquid chromatography* (pp. 319–341). Elsevier.

Hage, D. S., & Chen, J. (2005). Quantitative affinity chromatography: Practical aspects. In *Handbook of affinity chromatography* (pp. 615–648). CRC Press.

Katoch, R. (2011). *Analytical techniques in biochemistry and molecular biology.* Springer Science & Business Media.

Wilchek, M., & Chaiken, I. (2000). An overview of affinity chromatography. In *Affinity chromatography* (pp. 1–6). Humana Press.

To perform paper chromatography

29

Definition and rationale

Paper chromatography is a type of "partition" chromatography which means the stationary phase is a liquid or a liquid supported on an inert solid whereas the mobile phase is essentially a liquid or gas. The separation of different molecules is done according to the principle of differential portioning between the mobile and the stationary phase. In this chromatography technique, a highly purified cellulose filter paper is taken as a support and a stationary "liquid phase" is formed when water drops are settled in its pores. The fluid placed in a developing tank/jar consists of the mobile phase. Since both phases are in liquid state, paper chromatography can also be stated as "liquid-liquid" chromatography.

A mixture of products in the solvent can be separated based on its constituents if they have significantly differential portioning or different partition coefficients between the mobile phase and the stationary phase, i.e., water-saturated cellulose. The solute will tend to remain near the point of the application when the partition favors the aqueous phase and conversely the solute will move with the solvent flow when the mobile organic phase is favored.

Materials and equipment

- Chromatography chambers
- Trough assembly—glass troughs are preferred
- Drying chambers
- Heat source for color development (oven)
- Papers (Whatman No. 1)
- Sample application (10–50 µL)
- Solvent system

Step-by-step method details

1. In this method, a highly purified cellulose filter is taken and the solute mixture to be analyzed is placed at a marked spot or dot at the bottom of the paper, and this spot defines the origin.

Advanced Methods in Molecular Biology and Biotechnology. https://doi.org/10.1016/B978-0-12-824449-4.00029-3

2. The paper is then suspended in a suitable closed jar containing a solvent (mobile phase). It is then allowed to travel steadily with the help of capillary action along with the paper in the chromatography chamber.
3. Capillary draws the solvent through the paper, moves the components in the direction of the flow of the solvent, and dissolves the sample as it passes the origin dot. Since different components have different solubility in the mobile phase, they get separated from each other.
4. The solvent front is located and paper is dried and the positions of the compounds are detected by spraying the paper with indicating reagents like ninhydrin, etc.

Expected outcomes

The ratio of the distance moved by carbohydrate under consideration to the distance moved by the glucose (Fig. 27).

Quantification and statistical analysis

The different compounds are identified according to their R_f *(Relative to Front)* values, which is described as:

The ratio of the length/distance traveled by the compound from origin to the distance traveled by solvent from the origin. The R_f (Relative to Front) value is specific and constant for a particular compound and reflects the distribution coefficient for

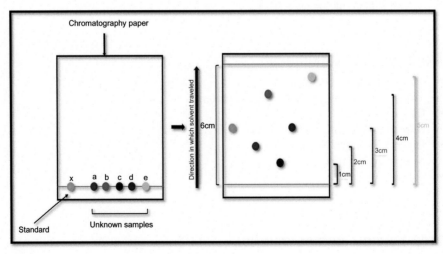

FIG. 27

Paper chromatography technique. Here, the spot "d" matches with the standard x. So "d" is x.

that compound under standard conditions. For example, the carbohydrates R_f value is defined as the ratio of the distance moved by carbohydrate under consideration to the distance moved by the glucose.

Advantages

It is a simple and rapid method and requires very less quantitative material. Compared to other analytical methods and equipment, paper chromatography does not occupy a large space. Its resolving power is excellent. It is used in DNA/RNA sequencing, separation of colored mixtures, e.g., pigments. It is used in the identification of unknown inorganic and organic compounds from a mixture of compounds.

Limitations

The application of a large number of samples cannot be done on paper chromatography. Paper chromatography is not effective in quantitative analysis. It cannot separate a complex mixture. It is not that much accurate as compared to HPLC or HPTLC.

Safety considerations and standards

Gloves should be worn during the experiment. The solvent used may release fumes, so the beaker should be covered with a plastic wrap when not in the hood.

Alternative methods/procedures

- Size-exclusion chromatography or gel filtration
- Thin-layer chromatography
- Ion-exchange chromatography
- Affinity chromatography

Further reading

Baum, C. (2009). *Genetic modification of hematopoietic stem cells*. Humana Press.

Block, R. J., Durrum, E. L., & Zweig, G. (2016). *A manual of paper chromatography and paper electrophoresis*. Elsevier.

Eigsti, N. W. (1967). Paper chromatography for student research. *The American Biology Teacher*, 29(2), 123–134.

Lederer, E., Lederer, M., & Chromatography, A. (1957). *Review of principles and applications*. *Vol. 2* (p. 711). New York: Elsevier.

Merritt, A. J. (1966). Paper chromatography. In I. M. Hais, & K. Macek (Eds.), *Vol. I. A comprehensive treatise*. New York: Academic Press. 1963. 955 pp. $26.50.

Growth of bacterial cultures and preparation of growth curve

30

Definition

Growth refers to the increase in the cell size number and mass during the development of an organism. The growth of the organism is affected by both physical and nutritional factors. The physical factors include the pH, temperature, osmotic pressure, hydrostatic pressure, and moisture content of the medium in which the organism is growing. The nutritional factors are also responsible for growth which include carbon, nitrogen, sulfur, phosphorous, and other trace elements provided in the growth medium.

- Bacteria are unicellular organisms and as the bacteria reach a certain size, they divide by binary fission, the growth process continues in a geometric fashion. After "n" number of divisions, the number of bacteria in a medium would be 2^n. The bacterium is then known to be in an *actively growing phase*.

Rationale

The bacterial growth population is studied when these cells of the bacterium are inoculated on the fresh broth followed by the incubation under controlled conditions. The bacteria used the nutrition of the media for an increase in its size and mass. To study the dynamics, a graph is plotted where (absorbance) versus the incubation time or log of cell number versus time is recorded. The curve thus obtained is a sigmoid curve and is known as a *standard growth curve*. The increase in the cell mass is measured by a Spectrophotometer which measures the turbidity and optical density. The degree of turbidity in the broth culture is directly related to the number of microorganisms present, either viable or dead cells. The amount of transmitted light through turbid broth decreases with a subsequent increase in the absorbance value.

The growth curve has four distinct phases:

1. *Lag phase*

   It refers to the pioneer stage of bacterial growth kinetics. When bacteria are introduced into the fresh medium, they take some time to adjust in the environment. This stage of growth is termed as the *Lag phase*. In this phase, bacterial cells do not divide but their metabolism and cell size increase. The previous growth conditions of the bacteria decide the length of the lag phase.

Advanced Methods in Molecular Biology and Biotechnology. https://doi.org/10.1016/B978-0-12-824449-4.00030-X

2. *Exponential or logarithmic (log) phase*

In this stage, the adjusted organisms in the freshly inoculated nutrient broth undergo rapid growth and divisions. Their cellular metabolism increases rapidly and they initiate DNA replication by binary fission at a constant rate. The nutrient medium is fully exploited and organisms divide at the maximal potential. The logarithmic (exponential) growth is achieved in this phase. A single cell divides into 2, which further replicates into 4, 8, 16, 32, and so on. This results in a balanced growth. The time taken by the microorganism to double its population varies from one organism to another and is known as the *generation time*. For example, *E. coli* has a generation time of 20 min, i.e., it divides after every 20 min, whereas for *Staphylococcus aureus*, it is 30 min.

3. *Stationary phase*

During the log phase, all the nutrients in the medium are exploited by the microorganism for their rapid multiplication and their growth continues. This leads to the accumulation of toxic metabolites and other wastes in the medium. Most of these have inhibitory properties and often change the physiochemical properties like pH, temperature, etc. of the medium. As a result, the reproduction slows down, and the cells undergoing division are equal to the number of cell deaths, and bacterial growth becomes stagnant. This stage is known as stationary phase. Though such a bacterial cell is inoculated in the fresh medium, it resumes the log phase.

4. *Decline or death phase*

Due to the continuous growth of microorganisms, the nutrient medium is depleted and the accumulation of metabolic waste products and toxic wastes in it continue to pile up. They lose their ability to reproduce and they start dying. This stage of the death of the bacterial population is technically called the death phase.

Materials and equipment

- Glassware
- Distilled water
- Spectrophotometer
- Micropipettes
- Shaking incubator
- LB broth

Step-by-step method details

- Prepare a fresh and well sterilized LB broth for bacterial inoculation.
- Then adjust the spectrophotometer and set the wavelength at 660 nm.
- Put 1 mL of uninoculated sterile media into a sterilized cuvette and blank the machine by setting it to 0 ABS with this sample. It is an important step to ensure further calculations and quantitate growth due to a change in turbidity as it allows us to standardize the turbidity of the media without any bacterial cells in it.

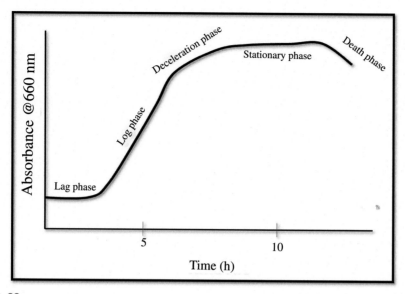

FIG. 28

Growth curve of *E. coli.*

- Now inoculate the primary media with 1 mL of overnight grown culture.
- After inoculating, immediately take 1 mL sample of the inoculated media and pipette it into a clean cuvette.
- Place it in the blanked spectrophotometer, and record the OD reading at time "0."
- After every 15 min repeat the previous step until there is no increase in absorbance.
- Plot all the readings on a graph.

Expected outcomes

A growth curve is obtained which shows the size of the bacterial population over the time.

The growth curve is hyperbolic due to the exponential bacterial growth pattern (Fig. 28).

Further reading

Baranyi, J., & Pin, C. (1999). Estimating bacterial growth parameters by means of detection times. *Applied and Environmental Microbiology, 65*(2), 732–736.

Gilpin, M. E., & Ayala, F. J. (1973). Global models of growth and competition. *Proceedings of the National Academy of Sciences of the United States of America, 70*(12), 3590–3593.

Ratkowsky, D. A., Lowry, R. K., McMeekin, T. A., Stokes, A. N., & Chandler, R. (1983). Model for bacterial culture growth rate throughout the entire biokinetic temperature range. *Journal of Bacteriology, 154*(3), 1222–1226.

Ratkowsky, D. A., Olley, J., McMeekin, T. A., & Ball, A. (1982). Relationship between temperature and growth rate of bacterial cultures. *Journal of Bacteriology, 149*(1), 1–5.

To study standard operating procedures (SOPs) for various equipment

Autoclave operation

Purpose

The main aim of this document is to provide standard operating procedures for the use of autoclaves. Autoclaving is a process that is used to kill microorganisms and decontaminate biohazardous waste and microbiological equipment used at biosafety levels 1–4.

Risk management

A. *Potential Risks:* High-pressure and high-temperature steam are used in autoclaves for sterilization. The possible safety risks for the operators include
- *Heat burns* from hot materials and from walls and door of the autoclave.
- *Steam burns* from residual steam that comes out from an autoclave and materials after the cycle completion.
- *Hot fluid scalds* due to boiling liquids and spillage in the autoclave and also during transport of heated materials.
- *Hand and arm* injuries while closing the door of an autoclave.
- *Body* injury in case of an explosion.

B. *Health and Safety:* To ensure the health and safety standards of the personnel using the autoclave, maintenance of autoclaves is very important for each and every department and to train personnel for its proper use.
- An individual shall be posted for the responsibility of the autoclave and his name should be labeled near the autoclave.
- SOPs should be posted outside of the autoclave.
- It is the supervisor's responsibility to make sure that all the employees are trained before operating any autoclave unit.
- Instructions and procedures provided by the manufacturer must be followed thoroughly.
- Personal protective equipment (PPE) and other protective clothing must be worn when loading and unloading the autoclave.

Advanced Methods in Molecular Biology and Biotechnology. https://doi.org/10.1016/B978-0-12-824449-4.00031-1

Laboratory centrifuge

Required controls for the use of centrifuges

- Appropriately trained persons should operate a centrifuge.
- Machine logbook must be completed (hours run determines the life of the rotor).
- Lid and seals must be checked for cleanliness and damage, before the use of the rotors. Rotors that are damaged must not be used.
- The tubes of the centrifuge should not be filled beyond the maximum limit as recommended by the manufacturer. The maximum speed stated must never be crossed.
- The rotor has to be balanced within the limits recommended (materials with similar densities must be in the opposite sides of the rotor).
- The centrifuge must be fitted and sealed in a particular place before operating it.
- The solvents that are being used should be in compatibility with the tube material. Properly fitting tubes must be used.
- Cleaning of spilled materiel must be done immediately.
- Flammable and highly explosive materials must be avoided.
- Lifting off the lid or slowing the rotor by hand must not be attempted especially when the rotor is in motion. Centrifuges must be repaired by authorized personnel.
- Faults must be reported immediately. Centrifuge must not be used until the faults are serviced and repaired. Never attempt the repair yourself.

Pre-run checks

- Each tube compartment should be free from contamination and corrosion.
- The rotor should be clear, free from any kind of corrosion, crack, and burrs around the rim. It should be noted that the centrifuge chamber, drive spindle, and the mountings are clean. The drive surfaces should be wiped before installing the rotor.
- The lid of the rotor should be secured and rotor to spindle securing device should be placed correctly before using the machine.

Hazards and risks associated

- Rotating parts usually face mechanical failure that can cause a violent hazard.
- Contact with rotating parts. Sample leaks cause aerosols, stress corrosion, and contamination.
- Any kind of imbalance in the samples may cause machine movement, fire, or explosion.

- Contact with the contaminated components or vapors may prove to be hazardous.
- Centrifuges are lethal, thus care and vigilance should be maintained all the times.

The following operating procedures must be followed:

Distillation unit

S. no.	Distillation column start-up SOP
1.	Set all controllers to manual
2.	Fill reboiler with liquid bottom product
3.	Open reflux valve and operate the column on full reflux
4.	Establish cooling water flow to condenser
5.	Start the reboiler heating coil power
6.	Wait for all of the temperatures to stabilize
7.	Start feed pump
8.	Activate reflux control and set reflux ratio
9.	Open bottom valve to collect product
10.	Wait for all the temperatures to stabilize

Electrophoresis equipment

There are many risks associated with operating the electrophoresis equipment. It involves the use of high voltage electricity, thus there is a high chance of electric shock. The chemicals involved in the experiments using electrophoresis are highly toxic.

- Before attempting the use of the equipment one should be fully trained by an experienced worker.
- Chemicals to be used should be studied well before use.
- The wiring and the equipment needs to be thoroughly checked for signs of damage before use. Use of worn out and frayed leads should be avoided.
- Electrophoresis tanks having secure design and materials should be used. High-quality equipment prevents contact with buffer when connected to a power supply.
- Power supply should always be disconnected before use and before moving the tank.
- Spills of the buffer or gel should be cleaned up immediately as these may contain highly toxic chemicals like EtBr or acrylamide.
- Nitrile gloves should be used instead of latex gloves while handling buffers and gels as the latter may contain pores in them.
- Care should be taken that the leakage from the upper buffer chamber do not cause arcing.

Emergency action in case of electric shock

- Switch off power at once. Do not attempt to touch the victim until they have been isolated from the source of electricity. If you are unable to get to the power supply you will have to insulate yourself so that you do not become a casualty. A piece of DRY wood or rolled up paper could normally be used.
- Check the casualty's airway and breathing, if unconscious be ready to resuscitate, call for help immediately.
- Treat any burns when away from the electricity by flooding the affected part with water for a minimum of 10 min. Remove and constrictions if swelling is likely, keep the injured part elevated.
- *Do not* burst blisters, apply any creams or sprays or apply any plasters. Seek help from a first aider.

Fridge and freezer
Process
Minimum PPDS

1. Safety glasses.
2. Rubber gloves.

Objectives
The proper use of the refrigerator/freezer to store chemicals and/or blood/serum.

Warnings
The refrigerators should never be used to keep edibles or food items.

The containers if broken or improperly closed may pose a respiratory threat to the workers because the vapors get concentrated in the closed spaces.

Operating procedure
When to store chemicals in refrigerators

- When required by the MSDS (Material Safety Datasheet) for that specific chemical.
- When a laboratory prepared chemical might undergo thermal decomposition or other undesired reaction.

When to store chemicals in the freezer

- Chemicals that rapidly degrade at room temperature or require very low temperatures to remain stable for any given period of time may be stored in the deep freeze.

General procedures

Proper labeling of stored chemicals

The chemicals that are kept in the refrigerators and deep freezers should always be labeled in the following manner

- The chemical name should be written exactly as it appears on the container. Structures or formulas alone are not sufficient. If the name is unknown (e.g., a photolysis mixture) state the reactants and what was done to them in general terms. For example, "A + B, photolyzed at 300 nm."
- If the compound is a mixture, properly list all components indicating their approximate percentages.
- Mention any kind of hazards associated with the chemical (i.e., carcinogenic, flammable, corrosive, etc.). Store carcinogens, suspected carcinogens, and stench compounds in a secondary container. A jar with a lid is most appropriate for stench compounds, while either a bucket or a jar is appropriate for carcinogens. This requirement for secondary containers for stench compounds is ABSOLUTELY SERIOUS.
- Proper reference notebooks must be maintained for the compounds or reaction mixtures prepared in the laboratory that includes the initials of the worker with notebook and page number. Date should also be mentioned.
- If the container is too small for labeling, place it into a beaker, try not to waste space by using an excessively large beaker, and label the beaker appropriately. Try to put a notebook reference on the container regardless of its size.
- Parafilm must be used to seal the containers for reduction of the odor.

Spills

- Spills should be immediately cleanedup following the appropriate procedure (see the Ames Lab Safety manual and the MSDS for the compound(s)). These may contain hazardous and toxic chemicals in them.

Development of frost in the freezer

- Door of the refrigerators should never be kept open for long periods of time as it may promote the development of frost in their freezer sections. Refrigerator and deep freezer doors should be closed as soon as possible, especially in the summer, to prevent frosting.

Maintenance

- A weekly inspection of the refrigerators should be performed by the supervisor. Any mislabeled containers should always be brought to the attention of the appropriate group member. Unlabeled chemicals should be disposed of properly.

- The need for defrosting should be assessed by the supervisor and will involve removal of the contents to another refrigerator or cooler and the following of defrosting procedures as stated by the manufacturer of the equipment.
- The reduction of odor should be performed by including an open pan or a glass container (temperature bath containers work well) of sodium bicarbonate in each refrigerator/deep freezer. The contents of each container should be disposed of properly on at least a monthly basis and replaced with fresh sodium bicarbonate.

Gel documentation system
Purpose
To provide instructions on the proper use of the gel documentation system (Gel Doc).

Safety precautions
- The workers before attempting to use the equipment must receive training from experienced and qualified persons. Supervisor should make sure that their personnel are well trained.
- EtBr is famously known to be a potential mutagen, thus lab coat and gloves should be worn before handling it and the stained agarose gels.
- UV light may cause damage to the eyes and skin, thus it should be avoided to look into the UV light source directly without any protection.

Notes
- If the gel needs to be moved from one room to another, it should be kept on a small tray or a secondary container, so that it can be transported without need for gloves. It is also permissible to use the "one-glove" technique described below:
- Gloves can be removed with the dominant hand while the other hand can be used to open and close the doors and carry the gel.
- The doors of the exposure chamber should be opened with the ungloved hand. Place the gel into the exposure chamber using the gloved hand and a spatula.
- Controls of the Gel Doc system should be manipulated with the ungloved hand.
- Signatures on the logbook should be made with the ungloved hand.
- After finishing the documentation process, doors should be opened with the ungloved hand; gel can be removed with the gloved hand and spatula.
- Wipe the glass surface of the Gel Doc system with tissue paper. Do this with the gloved hand. Use the same hand to get the gel back to the laboratory, opening all doors with the ungloved hand.

Procedure

- *Starting the program*
- Switch on the computer monitor using the ungloved hand
- Open the Gel Doc software and click on "File" from the menu bar, click on "acquire" and select "Gel Doc."
- *Setting up the gel*
- Open the door and load the gel in to the visualizing chamber, center the gel, and assist in visualization.
- Close the door and switch on the UV light.
- Using the ungloved hand, adjust the focus, zoom, and aperture on the camera to obtain the optimal image.
- In the Gel Doc window, click "Capture." Select the hatched-box icon in this window and drag it to select the area of interest.
- On the menu bar, select "Edit," and cursor down to "Extract." A new window will appear with the final picture. Image properties like brightness and contrast can be manipulated at this point.
- *Printing*
- Select "File" from the menu bar, go to the "video print" option and click to print.
- When the roll shows pink strip, get new roll and install as per the directions.
- Close the windows containing the extracted and original images. Click "Don't Save" in the pop-up dialog box.
- *Closing the program*
- Turn off the UV light.
- Using gloved hands, remove the gel from the chamber, wipe down the surfaces with a tissue paper or paper towel.
- Close the door to the chamber with the gloved hand.
- Record your name and the number of photos taken in the logbook.
- *Contingency Plans*
  - *Skin contact*
  - In case if the EtBr solution comes in contact with the skin, the area of the contact should be washed immediately with water and soap.
  - *Spills*
  - The area of the spill should be sealed off and assistance should be asked immediately.
  - Ensure that lab coat, shoes with closed toes, gloves, and eye protection should be worn.
  - If the spill has occurred on Benchkote, remove the affected section and place in a leakproof container or bag. Label the bag as ethidium bromide waste and submit a waste disposal requisition to EHS.
  - The gel if dropped on a hard surface, the liquid or gel should be adsorbed with paper towels or adsorbent pads.

- The used materials should be placed in leakproof bags and labeled as EtBr waste.
- The solution of sodium nitrite and hypophosphorous acid can be used to decontaminate the area of spillage. Add 4.2-g sodium nitrite and 20 mL of hypophosphorous acid to 300 mL of water.
- UV light may be used to locate any remaining ethidium bromide.
- Soak a paper towel with the decontamination solution and wash the affected area.
- Water is used to rinse the area few more times to ensure the cleanup of residue. Paper towels are used to clean the area again.
- Used materials should be put in the ethidium bromide waste and a chemical waste disposal requisition should be submitted to EHS.

Incubator/shaker
Required controls for the use of incubators

- Label the containers and materials placed in the incubators with name, date, and contents.
- Remove the containers and materials after the required incubation time to prevent overcrowding or contamination.
- Report immediately to the laboratory technician if any kind of spillage occurs in the incubator.
- Wear the proper kind of gloves when the temperatures in excess of 37°C are in use.
- Report the faults associated with the incubator. If the faults are observed, switch it off until repaired.
- Unauthorized/untrained personnel should not attempt to dismantle any part of an incubator for any purpose.
- Report any kind of cuts or burns. Read the instructions carefully.

Hazards associated with the use of incubators

- Heat, electricity, chemicals, microorganisms, damaged containers.

Risks associated with the use of incubators

- Microbial contamination of body and clothing
- Burns to skin
- Electrical shock
- Chemical contamination of body and clothing

People at risk when using incubators

- Students, academic staff, technicians

Thermocycler/PCR
General

The thermal cyclers contain heating blocks programmable to cycle at required temperatures over required time periods. First-use training in the use of the machines is required, to avoid accidents, misuse, and subsequent shortening of the life of the machines. Read this SOP, and the relevant equipment manual for the particular thermal cycler to be used (located on the shelf above the lab manager's desk).

Hazards

Operation of this machine is considered low risk.

- The heating block can reach high temperatures, thus posing a risk of minor burns.
- Slight risk of injury from moving parts (squashing fingers in heating block).
- General electrical problem if the machine is faulty, i.e., shorting.

Risk control measures

- Training is required in a proper way and care needs to be taken to use of the thermal cyclers before the first use.
- Contact time of the heating block should be limited.
- Maintenance checks should be made to ensure if the machine is electrically safe and functioning properly.

General points
Hours of use

All hours.
   Authorized users:
   Prepare your samples.

- Put your samples into the heating block of the thermal cycler and gently close the lid of the heating block.
- The switch on the back of the machine should be turned on and wait for it to run through the self-test of the heat pump.
- Equipment manual should be referred for the details of operating the thermal cycle.

- Enter your program, or locate and find your program in the list, and then begin the cycling program.
- Remove your samples when the cycling is complete, and turn off the machine.

Notes

If you book a thermal cycler, and subsequently decide that you do not need it, please cross out your booking, so that other people realize that the machine is available for use.

DO NOT leave the heating blocks set to 4°C for long periods of time (e.g., overnight), it reduces the life of the machine as it has to work harder to maintain that temperature. Your samples will be safe at room temperature overnight, or at least set the temperature to 15°C or above.

Waste management

Any unwanted hazardous materials must be disposed of in an appropriate manner.

Regular monitoring and/or maintenance procedures

Ensure the heating block and lid are clean and air vents are free of dust buildup.
Regular electrical check (organized by the lab manager).

Pipette

1	Clean micropipette should be used. The same tip can be reused if re-pipetting from the same solution
2	Micropipette tip should be firmly attached with the end of the tip
3	Fingers should be kept away from the pipetting tip as human skin contains RNase and oils. If the pipetting tip gets touched, that part should not be immersed in the solution
4	Pipette should be filled up slowly rather than snapping back the filling plunger which can cause the air bubbles to form in the sample and sample may get adhered onto the upper part of the tip as separate drops which are very hard to transfer. Note: viscous samples take much longer to fill (and discharge)
5	The tip should not carry any extra drop after filling it from the solution. It is advisable to gently touch the tip to the inside of the container to remove any extra drop.
6	Be able to feel the difference in spring pressure between the fill and the blow-out positions of the plunger
7	Samples should be discharged slowly. Doing it quickly may leave some of the sample or solution in the pipette. This is especially true for viscous samples

Continued

8	Be sure that the last part of a drop is transferred. It is advisable to touch the tip to the inside of the vessel to ensure that the last part of a drop is transferred
9	Worker should be aware of the volume adjustments and should read the calibration scales of the variable micropipettes
10	Check the calibration of the micropipette by weighing (on a water-resistant tare; on a microbalance) a measured volume of water. (The density of water at 20°C is 0.9982071 g/mL. Be aware that the weight of the water drop will change as the water evaporates from the tare.)
11	Cleanup when spilled (wipe with DI water)

Pipetting instructions
Macropipettes

Ask for help if you need help

1	Mouth should not be used for pipetting
2	Use an appropriately sized pipette and pro-pipettor
3	Do not transfer an "extra" drop! (touch the tip of the pipette to the side of the container when filling)
4	Do not forget to transfer the "last" drop! (touch the tip of the pipette to the side of the container when emptying)

Micropipettes: Before using the micropipettes, read the instructions for their proper use:

Select the right micropipettes

Volume range	Volume	Volume increment	Inaccuracy	Imprecision
Dark Gray operating button, for use with 10-μL pipette tips				
0.1–2.5 μL	0.25 μL	0.002 μL	±12.0%	≤6.0%
	1.25 μL		±2.5%	≤1.5%
	2.5 μL		±1.4%	≤0.7%
Gray operating button, for use with 20-μL pipette tips				
0.5–10 μL	1 μL	0.01 μL	±2.5%	≤1.8%
	5 μL		±1.5%	≤0.8%
	10 μL		±1.0%	≤0.4%
Yellow operating button, for use with 200- and 300-μL pipette tips				
2–20 μL	2 μL	0.02 μL	±5.0%	≤1.5%
	10 μL		±1.2%	≤0.6%
	20 μL		±1.0%	≤0.3%
10–100 μL	10 μL	0.1 μL	±3.0%	≤1.0%
	50 μL		±1.0%	≤0.3%
	100 μL		±0.8%	≤0.2%

Continued

Volume range	Volume	Volume increment	Inaccuracy	Imprecision
20–200 µL	20 µL	0.2 µL	±2.5%	≤0.7%
	100 µL		±1.0%	≤0.3%
	200 µL		±0.6%	≤0.2%
Blue operating button, for use with 1000-µL pipette tips				
100–1000 µL	100 µL	0.1 µL	±3.0%	≤0.6%
	500 µL		±1.0%	≤0.2%
	1000 µL		±0.6%	≤0.2%
Violet operating button, for use with 5-mL pipette tips				
500–5000 µL	500 µL	5 µL	±2.4%	≤0.6%
	2500 µL		±1.2%	≤0.25%
	5000 µL		±0.6%	≤0.15%
Turquoise operating button, for use with 10-mL pipette tips				
1–10 mL	1 mL	10 µL	±3.0%	≤0.6%
	5 mL		±0.8%	≤0.2%
	10 mL		±0.6%	≤0.15%

How to adjust micropipettes

500–5000 µL	100–1000 µL	10–100 µL	0.5–10 µL
3	0	0	0
7	6	8	7
5	5	7	2
0	2	0	0
3.750 mL	0.652 mL	0.087 mL	0.0072 mL

Note: 1 mL = 1000 µL.

Water bath

Objective

To ensure that the instrument performs satisfactorily and gives accurate and reproducible results.

Scope

This SOP covers the operating procedure of water bath and this SOP is applicable to Quality Control Department.

Procedure

- *Procedure for General Cleaning*
- The power supply to the instrument should be switched off before cleaning it.
- For cleaning purpose, a dry cloth should be used and cleaning should be done every day.
- Adhered dust should be removed by wet mopping with a detergent. After this, the surface should be wiped with a clean dry cloth so that there are no traces of detergent or water on it.
- Water in the water bath should be changed every 2 days and use always demineralized water.
- *Operating Procedure*
- Sufficient amount of water should be available in the water bath.
- Instrument should be properly connected to the power supply.
- Fill the bath to the level little above the false bottom
- Switch on the supply. Put the toggle switch ON.
- Set the thermostat knob to the required temperature and the rise in the temperature is indicated by the glow of the RED INDICATOR LAMP.
- When the required temperature is reached, the RED INDICATOR LAMP will go OFF.
- The bath is then allowed to work.
- After completion of the heating process, the temperature is reduced to zero, by rotating the temperature-setting knob to zero position. Power supply is switched off and the bath is allowed to cool to the required temperature before the control knob is turned backwards. This is to eliminate unnecessary strain on the thermostat.
- Any discrepancy if observed during operation of the instrument should be reported to the department head notifying the defect to the Utility Department to rectify the defect. Affix "BREAK DOWN" label on the instrument till it gets rectified.

Further reading

Ng, Y. S., & Srinivasan, R. (2009). An adjoined multi-model approach for monitoring batch and transient operations. *Computers & Chemical Engineering, 33*(4), 887–902.

Wesselschmidt, R. L., & Schwartz, P. H. (2011). The stem cell laboratory: Design, equipment, and oversight. In *Human pluripotent stem cells* (pp. 3–13). Humana Press.

Index

Note: Page numbers followed by *f* indicate figures and *t* indicate tables.